D0960246

R. MCNEILL ALEXANDER

Dynamics of Dinosaurs and Other Extinct Giants

Columbia University Press
NEW YORK

COLUMBIA UNIVERSITY PRESS
New York Guildford, Surrey

LIBRARY OF CONGRESS
Library of Congress Cataloging-in-Publication Data

Alexander, R. McNeill.
Dynamics of dinosaurs and other extinct giants / R. McNeill Alexander.
p. cm.
Includes bibliographies and index.
ISBN 0-231-06666-X
1. Dinosaurs—Locomotion.
2. Extinct animals—Locomotion.
3. Animal mechanics.
I. Title.
QE862.D5A33 1989
567.9'1—dc19 88-20373
CIP

Printed in the United States of America

Casebound editions of Columbia University Press books are Smyth-sewn
and printed on permanent and durable acid-free paper

CONTENTS

PREFACE

THIS IS a book about big, extinct animals. Most of it is about dinosaurs, including the biggest land animals that have ever lived. There is a chapter about the huge reptiles that lived in the sea in the time of the dinosaurs and one about the biggest of the flying reptiles that lived at the same time. There are also short chapters about gigantic, extinct birds and mammals. Some very big reptiles lived before the dinosaurs but I have written nothing about them, because all of them were much smaller than the biggest dinosaurs and because fewer people are interested in them.

There are many other books about dinosaurs but none, I think, like this one. I have used the methods of physics and engineering to try to discover how extinct animals could have lived and moved. I like to think about animals in the kinds of ways that engineers think about machines and vehicles. That seems the best way of finding out how dinosaurs could have worked.

You do not need to know much science to understand this book. I have tried always to start from basics, and to keep the arguments simple. However, if you do already know a lot of science, please do not be put off by the simple explanations, because I think you will find ideas here that you have not read elsewhere. This book is meant for everyone who is interested in dinosaurs: scientists and nonscientists, schoolchildren and professors.

1

Introducing the Dinosaurs

THIS CHAPTER tells what dinosaur fossils are like and how they are formed and introduces some of the best-known dinosaurs. It is an introductory chapter intended mainly for readers who have little previous knowledge of these remarkable reptiles. Other may prefer to skip quickly through the chapter or to go directly to chapter 2.

Figure 1.1 shows a dinosaur 13 meters (43 feet) tall, well over twice the height of a full-grown (5.5 meter) giraffe. It is far bigger than any modern land animal and its skeleton (now in Berlin) is one of the world's most impressive museum exhibits.

This dinosaur, *Brachiosaurus,* is the largest known with a reasonably complete skeleton, but may not be the largest that ever lived. Possible rivals such as *Supersaurus* and *Ultrasaurus* are known by only a few bones each, and scientists are still arguing about the sizes of the complete dinosaurs.

This book is about gigantic animals, so most of the dinosaurs in it are very big ones, but figure 1.2 shows that there were smaller ones too. The *Compsognathus* was about the size of a hen and the *Psittacosaurus* was smaller than a pigeon, but both of them were juveniles. The *Compsognathus* skeleton (including the tail) is 0.8 meters long, which is not too much less than the 1.4 meters of another specimen that is probably adult. The *Psittacosaurus,* however, belongs to a species that grew to an adult length of 2 meters, which is still small for a dinosaur.

Figures 1.1 and 1.2 show dinosaurs as they are thought to have looked in life, but the actual specimens are mere skeletons. Figure 1.3 shows one of the fossils of *Psittacosaurus* (an adult). Some of the leg and tail bones are missing but have been drawn in, as they are thought to have looked, with dotted lines. The skeleton has collapsed or been squashed after death, making ribs stick out at odd angles, but the bones have stayed together. Many other dinosaur finds have been of scattered or jumbled bones.

Many big dinosaur skeletons have been separated bone by bone from the rock and reassembled as complete skeletons in natural postures. The *Psittacosaurus* skeleton (figure 1.3) was given much less drastic treatment. Only enough rock was removed to reveal all the bones, which were left attached to the remaining block of stone. The skeleton in natural posture (figure 1.4) was not actually constructed but the drawing was built up from drawings of the individual bones. Making drawings like this is an important stage in reconstructing the appearance of extinct animals. Adding flesh and skin to make drawings like figure 1.2 depends largely on knowledge of the soft anatomy of living animals, but also requires a good deal of guesswork.

Fossils get preserved because the earth's surface is continually changing. Massive earth movements crinkle it up into chains of mountains, which get worn down again by the processes of erosion. The Rockies, the Andes and the Alps are all more recent than the dinosaurs, and older mountains, such as the Appalachians, are less impressive because there has been more time for them to be eroded. The next few paragraphs explain how mountain building and erosion preserve fossils and make them possible to find.

Exposed surfaces of rocks get heated by the sun during the day and

FIGURE 1.1. *Brachiosaurus* with a house and an adult giraffe drawn to the same scale. The dinosaur, which is 13 meters tall, is drawn from a model sold by the Natural History Museum, London.

FIGURE 1.2. Juvenile *Compsognathus* (behind) and *Psittacosaurus* (right), with a domestic pigeon. The dinosaur skeletons on which the drawings are based are among the smallest known. Drawn by Matthew Hyman. From Coombs 1980. Reprinted by permission. Copyright © 1980 Macmillan Magazines Ltd.

cool during the night. This makes them expand and contract, breaking fragments off. Water seeps into cracks in rocks and freezes in winter, swelling and splitting the rock. Sand grains carried along by streams scour away the rock of the stream bed. These processes break rocks down into sand grains and mud particles which are washed away by rainwater and streams, which carry them down toward the sea. The sand and mud settle out where the water runs more slowly, on flooded plains or on sand banks or mud flats at a river's mouth. Carcasses of animals that die in these places may get buried in the sand or mud (figure 1.5a–c). Carcasses also get buried in sand dunes: that is what happened to the *Psittacosaurus,* which seem to have lived in a dry, sandy area. Yet other carcasses get buried in the calcium carbonate which, in certain circumstances, precipitates out in the sea. I do not know of any dinosaur fossils that got preserved like that but many of the fossils of marine reptiles, described in chapter 9, are in limestones formed from calcium carbonate precipitates. The flesh of carcasses rots away but buried skeletons may be preserved. Skeletons that do not get buried are generally destroyed, broken down by the same weathering processes as erode rocks.

As time passes and more sediment accumulates, skeletons get buried deeper and deeper (figure 1.5d). The sediments consolidate into rock:

sand becomes sandstone, mud becomes shale and calcium carbonate precipitates become limestone. The fossil may come to lie in solid rock, deep below the ground. If it stayed there it would never be found unless perhaps by mining or quarrying. However, earth movements crinkling the earth's surface may raise it up in a mountain (figure 1.5e) and erosion, wearing the mountain away, may expose it (figure 1.5f). If those things happen it may be found, but some of its bones may get broken or worn away first. The *Psittacosaurus* in figure 1.3 had lost parts of its feet by erosion, before it was found.

Usually only skeletons survive as fossils, but there are sometimes traces of other parts. If a carcass gets buried before the soft parts decay, the sediment will mould itself to the skin surface and in rare cases this impression of the skin may survive. Impressions of skin have been found with several fossils of duck-billed dinosaurs (dinosaurs like the one shown in figure 1.11). They show that the skin was scaly like tortoise or lizard skin. Fossil stomach contents are sometimes found: for example, one of the *Compsognathus* fossils has a lizard skeleton inside it. There are also some fossil dinosaur eggshells, which resemble the eggshells of birds and modern reptiles. The most famous belong to *Protoceratops* (a close relative of *Psittacosaurus*) as fossil embryos found with them show. Adult *Protoceratops* were 2 meters long and the eggs measure 10–20 centimeters. They seem to have been laid in hollows in the ground in clutches of thirty or more. Finally, there are a lot of dinosaur footprints: they will be described in chapter 3.

Figure 1.5 shows two dinosaur skeletons getting buried, one after the other. By the time the small one died (figure 1.5c) the big one was

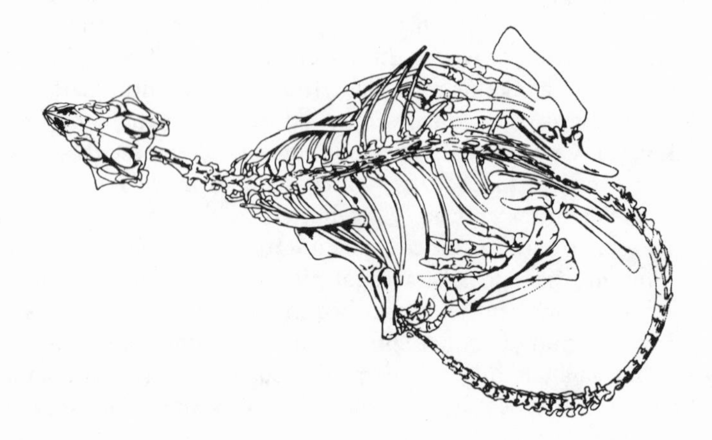

FIGURE 1.3. A *Psittacosaurus* skeleton as it was found, with enough rock cleared away to show the bones. From Osborn 1924.

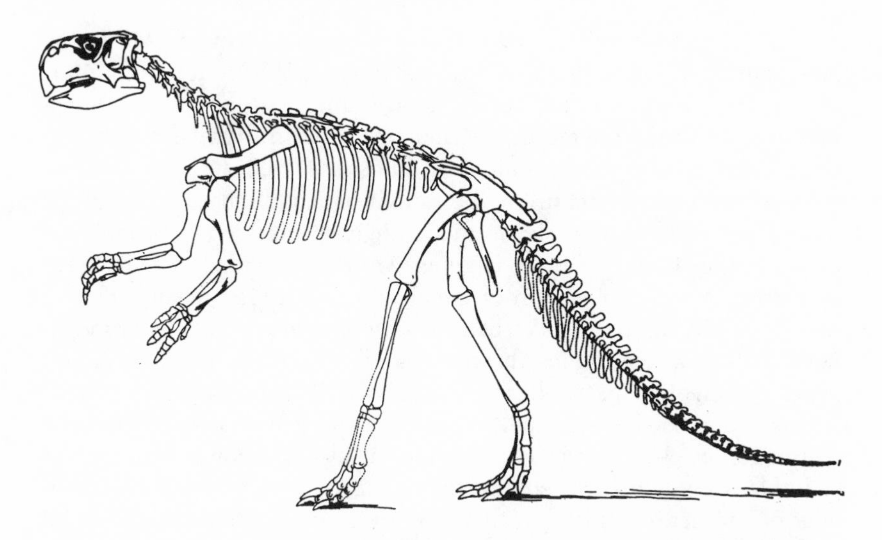

FIGURE 1.4. A reconstruction of the skeleton of *Psittacosaurus,* based on the skeleton shown in figure 1.3. The overall length of the skeleton was 1.4 meters. From Osborn 1924.

already buried, so the small one finished up in a higher layer of rock. It would be obvious to a paleontologist who found both fossils that the big one was the earlier of the two.

The relative ages of fossils are less obvious when they are found far apart, possibly in different continents, but they can be worked out by an extension of the same principle. Layers of rock of the same age all over the world can be matched up by the similarity of their fossils. Nineteenth-century geologists discovered this and used it to divide time up into four eras: Precambrian (the oldest), Palaeozoic, Mesozoic, and Cenozoic (table 1.1). Fossils are rare in Precambrian rocks but are plentiful in rocks from the later eras. All the dinosaurs lived in the Mesozoic era. Each era is subdivided into periods. The Mesozoic consists of the Triassic, Jurassic, and Cretaceous periods, with dinosaurs living in all three.

The periods were given names before anyone could tell how long each of them had lasted, but the discovery of radioactivity made dating possible. Radioactive isotopes break down into other isotopes, each at its own characteristic rate. Some break down exceedingly slowly, over periods of hundred of millions of years. When they are found in rocks it is sometimes possible to calculate the rock's age from the proportion of original isotope to breakdown products. Such measurements tell us that the Mesozoic era began about 230 million years ago and ended 65 million years ago. The dates are shown in table 1.1.

Each dinosaur species lived for only a small part of the Mesozoic. For example, *Compsognathus longipes* (figure 1.2) lived late in the Jurassic period, about 140 million years ago, and *Psittacosaurus mongoliensis* lived in the Cretaceous period, about 90 million years ago. Table 1.2 shows when these and other dinosaurs lived.

I gave each species its full name in that paragraph, because I wanted to emphasize that I was talking about single species. Every animal species, whether living or fossil, is given two names, which usually have meanings derived from Latin or Greek. *Psittacosaurus* means "parrot-lizard" (notice its parrot-like beak) and *mongoliensis* means "mongolian" (telling where the fossils were found), so the two names together mean "mongolian parrot-lizard." Similarly *Compsognathus longipes* means "long-footed pretty jaw." The first name in each case is the name of the genus, which may include several closely related species (for example lions *Panthera leo* and tigers *Panthera tigris* are both members of the genus *Panthera*). If there are several species, the second name tells us which is meant. It so happens that only one species of *Psittacosaurus* is known, and one of *Compsognathus*, so it will generally be unnecessary to use the second names until more species are discovered. However, several species of *Brachiosaurus* have been recognized, including *Brachiosaurus altithorax* from Colorado and *Brachiosaurus brancai* from Tanzania. Even in cases like this it is unnecessary to use the second name if it does not matter which of the closely similar species is being referred to. Most of the things that I might

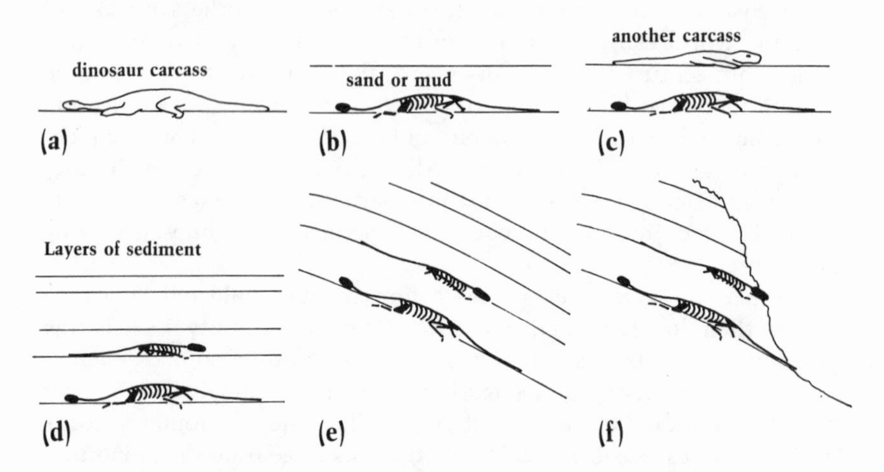

FIGURE 1.5. Diagrams showing how dinosaurs have been fossilized, and how fossils are brought to the surface by earth movements and erosion. The sequence of events is described in the text.

want to write about *Brachiosaurus brancai* (the best-known species) would also be true of *Brachiosaurus altithorax*.

There are many kinds of reptiles, and table 1.3 shows how they are classified. The dinosaurs are put in the group called the Archosauria, which also includes the crocodiles. The members of this group can be recognized by their teeth and by the holes in their skulls. Their teeth are set in sockets, not just stuck to the jaw like other reptile teeth. There are more holes in the sides of their skulls than in any other group of reptiles (figure 1.6). Holes 1 and 2 (for the nostril and eye) can be found in the skulls of all reptiles. They are the only holes in the sides of the skulls of turtles and other Anapsida, but other reptiles have one or both of holes 3 and 4 behind the eye. Hole 5 in front of the eye is peculiar to archosaurs, but crocodiles do not have it. In addition crocodiles and some other archosaurs have hole 6 in the lower jaw.

Few people would confuse crocodiles with dinosaurs: crocodiles (including alligators, etc.) are crocodile-shaped, and they have many distinctive anatomical features. Pterosaurs, the other well known archosaurs, are also obviously different from dinosaurs: they are the winged reptiles that are the subject of chapter 8. There were also some early archosaurs called thecodonts, which are less obviously different from the smaller dinosaurs, but the joints of the legs show an important difference. Thecodont leg joints show that they must have walked like crocodiles, with their feet well out on either side of the body. Dinosaurs walked like birds and mammals, with their feet well under the body (figure 3.5).

Now I will review the main groups of dinosaurs, introducing most of the genera that will feature in later chapters. I will describe their

TABLE 1.1. The divisions of geological time.

Era	*Period*	*Date of Beginning (million years ago)*
Cenozoic	Quaternary	2
	Tertiary	65
Mesozoic	Cretaceous	140
	Jurassic	190
	Triassic	230
Palaeozoic	(six periods)	570
Precambrian		

NOTE: Dinosaurs lived during the Mesozoic era.

appearance, making particular mention of their teeth and other evidence of what they ate. The dinosaur groups and the generic names of examples are shown in table 1.2.

The group of dinosaurs called the theropods seem, from the shapes of their teeth, to have been flesh eaters. This sets them apart from all the other dinosaurs, which ate plants. The theropods had relatively small fore legs and presumably walked on their hind legs. *Compsognathus* (figure 1.2) was one of the smallest of them. Its small, sharp teeth look suitable for eating insects and small vertebrates, and I have already mentioned a lizard found in one as fossil stomach contents. It lived late in the Jurassic but there had been similar small theropods since the Triassic. The ancestors of all the dinosaurs were probably rather like *Compsognathus*.

Allosaurus (figure 1.7) lived at the same time but was enormously larger. Its big teeth had serrated edges like steak knives, and look suitable for slicing through flesh. It probably attacked large plant-eating dinosaurs: in a later paragraph I describe one that seems to have been eaten by it.

Tyrannosaurus (figure 1.8) is a very late dinosaur, from the end of the Cretaceous. It is the biggest known flesh-eating dinosaur, with 15 centimeter steak-knife teeth. It has big hind legs and small front ones like other theropods, only more so. Its tiny front legs look useless.

TABLE 1.2. Classification of dinosaurs mentioned in this book and the periods in which they lived.

	Late Jurassic	*Early Cretaceous*	*Late Cretaceous*
SAURISCHIANS			
theropods	*Allosaurus* *Compsognathus*		*Tyrannosaurus*
sauropods	*Brachiosaurus* *Diplodocus* *Apatosaurus*	*Brachiosaurus*	
ORNITHISCHIANS			
ornithopods		*Iguanodon*	*Anatosaurus* *Parasaurolophus*
ceratopians		*Psittacosaurus* *Protoceratops*	*Triceratops*
pachycephalosaurs			*Stegoceras*
stegosaurs	*Stegosaurus*		
ankylosaurs			*Euoplocephalus*

The next major group, the sauropods, includes the largest dinosaurs. The best known ones lived late in the Jurassic period. *Diplodocus* (figure 1.9) was extraordinarily long (27 meters) but was much skinnier and presumably much lighter than *Brachiosaurus* and the other giants that have already been mentioned. It had a very long neck and an exceedingly long tail (I discuss their functions in chapter 5), with a relatively short body between. Its head looks small compared to the rest of the animal but is about the size of a rhinoceros head. Across the front of each jaw it has a row of tall thin teeth, like the teeth of a huge comb, which look suitable for plucking leaves and shoots from trees and other plants. It cannot have eaten flowering plants (which did not evolve until the Cretaceous) and probably ate the leaves of conifers, the commonest trees of its time. Another sauropod has been found with fossil stomach contents which consist (rather surprisingly) of fragments of woody twigs of about one centimeter diameter. Any leaves that were eaten with the twigs have been digested or decayed.

Large leaf-eating mammals such as giraffes pluck leaves with their tongues and front teeth, and grind them with their back teeth, but *Diplodocus* has no back teeth and its front ones seem unsuitable for grinding food. It has been suggested that sauropods swallowed stones and kept them in their stomachs. Movements of a muscular stomach wall could have rubbed the stones together, grinding any food that was in the stomach at the time. Plant-eating birds hold stones in their gizzards and use them in this way: for example, ostriches have up to 900 grams (two pounds) of pebbles.

Apatosaurus (figure 1.7) was shorter but stouter than *Diplodocus*. There used to be confusion about its name, because a fossil named *Apatosaurus* and another originally named *Brontosaurus* turned out to be identical. *Apatosaurus* is accepted as the correct scientific name, but people still speak informally of brontosaurs. One fossil of *Apato-*

TABLE 1.3. How reptiles are classified.

class Reptilia	
subclass Anapsida	earliest reptiles, turtles, etc.
subclass Lepidosauria	lizards, snakes etc.
subclass Archosauria	thecodonts
	crocodiles
	dinosaurs
	pterosaurs (ch. 8)
subclass Euryapsida	plesiosaurs (ch. 9). etc.
subclass Ichthyopterygia	ichthyosaurs (ch. 9)
subclass Synapsida	the ancestors of mammals

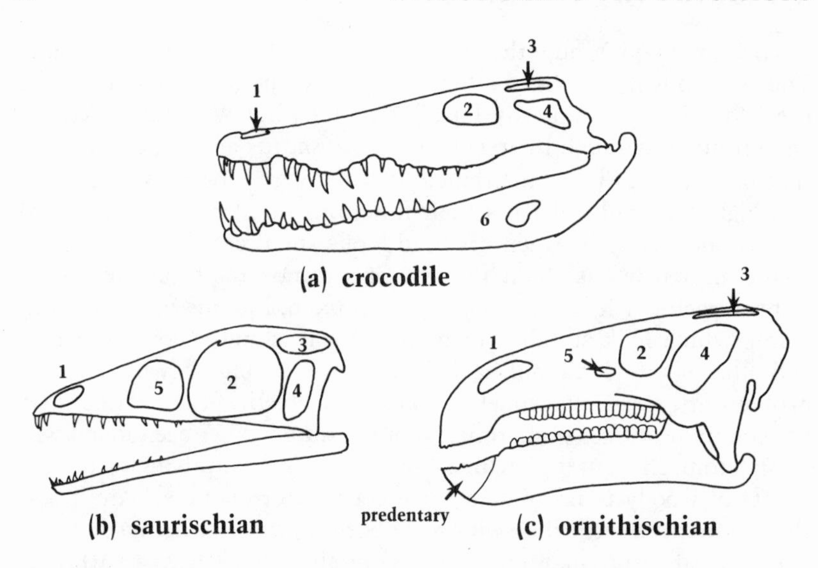

FIGURE 1.6. Skulls of (a) a crocodile; (b) a saurischian dinosaur (*Compsognathus*); and (c) an ornithischian dinosaur (*Iguanodon*). Skulls (a) and (c) have been distorted to show hole 3, which is in the roof of the skull and would be hidden in a true side view.

saurus has parallel scratches on its ribs, spaced about the same distance apart as the teeth of *Allosaurus*. Beside it was found a broken-off *Allosaurus* tooth. It seems that an *Allosaurus* fed on the carcass, but there is nothing to show whether it killed the sauropod (which was much larger than it could have been) or found the sauropod already dead.

Sauropods had elephant-like fore feet that would not have been much use for handling things. It seems obvious that they walked on all fours (unlike theropods), and there are fossil footprints that seem to confirm this. (They are described in chapter 3). However, the shortness of the fore legs of *Diplodocus* and *Apatosaurus* suggests that they may have evolved from bipedal ancestors. Also, there are early members of the

FIGURE 1.7. *Apatosaurus, Allosaurus* (the biped), and a large African elephant.

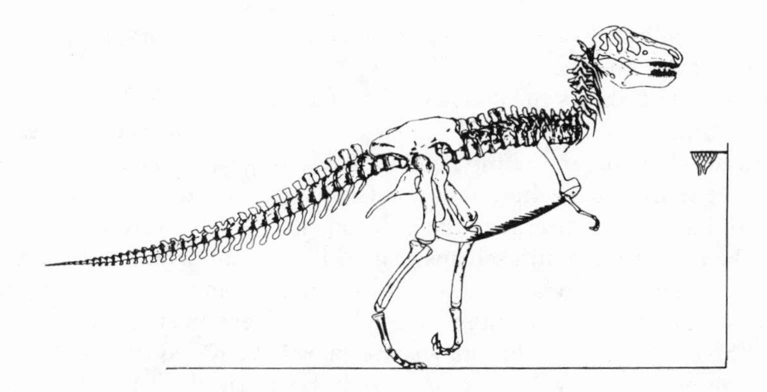

FIGURE 1.8. *Tyrannosaurus,* from Newman 1970, and a basketball net. The mouth of the net is 3.05 meters (10 ft) from the floor.

group that may have been bipeds. *Brachiosaurus* (figure 1.1) was a differently proportioned sauropod, with very long front legs and shorter hind ones. Its vertebrae fit together best with the neck nearly vertical, suggesting that that was how the neck was carried. The long front legs and near-vertical neck would have enabled it to reach high branches, to feed like a giraffe. Its teeth were broader than *Diplodocus'* teeth but still seem suitable only for plucking off leaves.

The theropods and sauropods together form the Saurischia, one of the two main groups of dinosaurs. The remaining dinosaurs (all of them plant eaters) form the Ornithischia. The most obvious differences between saurischians and ornithischians are in their jaws and hips. Compare the skull of *Iguanodon* (an ornithischian, figure 1.6c) with that of *Compsognathus* (a saurischian, figure 1.6b). The ornithischians had no front teeth and presumably had horny beaks, like the beaks of birds and turtles, on the fronts of their jaws. (I have already mentioned the parrot-like beak of *Psittacosaurus,* another ornithischian.) Another pe-

FIGURE 1.9. *Diplodocus* and a San Francisco cable car. The dinosaur, which is 25 meters long, is drawn from a model sold by the Natural History Museum, London.

culiarity of ornithischians is an extra bone (the predentary) at the front of the lower jaw.

Figure 1.10 shows the pelvic girdles of *Tyrannosaurus* (a saurischian) and *Psittacosaurus* (an ornithischian). You can see how these bones fit into the skeleton, in the hip region, by referring back to figures 1.8 and 1.4. In saurischians the pubis points forward and the ischium backward, but in ornithischians the pubis also has a backward extension.

The first of the ornithischians that I will describe belong to the group called the ornithopods. *Anatosaurus* (figure 1.11) and *Iguanodon* (figure 3.6) are examples. They must have carried at least most of their weight on their big hind legs, but may sometimes have rested their front feet on the ground as well. *Anatosaurus* has a broad duck-like beak but *Iguanodon* has a deeper, narrower one. Both have impressive batteries of grinding teeth (a striking difference from sauropods) and must have chewed their food like cattle. It has been suggested that they may have had cheeks enclosing the sides of their mouths to prevent half-chewed food from falling out. Mammals have cheeks, but modern reptiles do not.

Teeth that were simply pressed together would crush food, but to grind food they must slide over each other. Horses and cattle grind by moving their lower jaws from side to side, but ornithopod jaws worked differently. The lower jaws moved straight up and down but the upper jaws were hinged along their upper edges and splayed apart when the teeth pressed together (figure 1.12). Thus the teeth slid over each other, grinding the food rather than merely crushing it. No one has seen this mechanism working (there is nothing like it in living animals), but

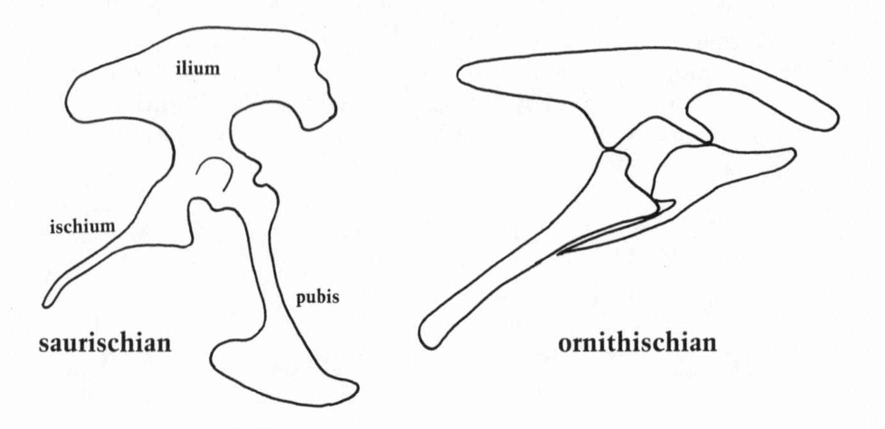

FIGURE 1.10. Pelvic girdles of a saurischian (*Tyrannosaurus*) and an ornithischian (*Psittacosaurus*).

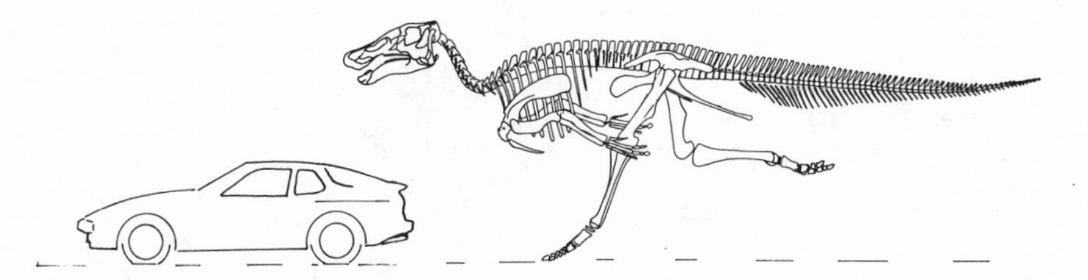

FIGURE 1.11. *Anatosaurus,* from Galton 1970, chasing a Porsche. The dinosaur is 8.9 meters long and the car 4.3 meters.

study of the structure of the jaws and their joints suggests that they must have worked in this way. Further evidence has been obtained by looking at the worn surfaces of the teeth through a microscope: the fine scratches on them run across the teeth, as they should if the jaws moved as suggested.

The ceratopians or horned dinosaurs were another important group. *Psittacosaurus* (figure 1.4) is a ceratopian, but it is unusual in two ways: it had no horns, and it seems to have walked on two legs whereas other ceratopians walked on all four. *Triceratops* (figure 1.13) is more typical, and is one of the largest ceratopians. It has two long horns, one over each eye, and a short one on its snout. It also has a great frill of bone extending back from the skull over the neck. Some other ceratopians have even longer frills. I discuss these horns and frills in chapter 6.

Ceratopians had impressive batteries of back teeth, arranged differ-

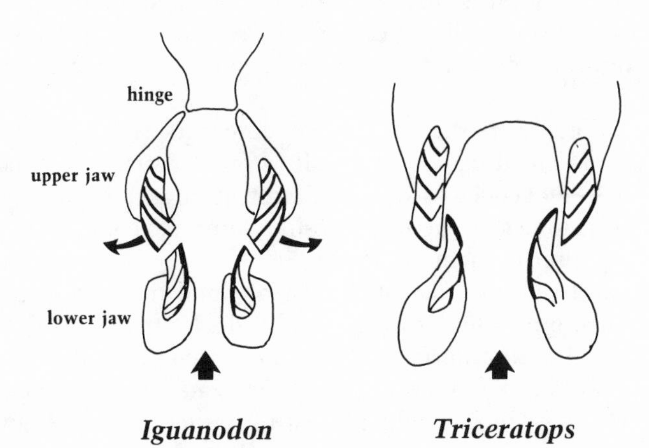

FIGURE 1.12. Diagrammatic sections through the jaws of an ornithopod and a ceratopian. Enamel coatings on tooth surfaces are represented by thick lines. Arrows show how the jaws moved.

FIGURE 1.13. *Triceratops* (center) and *Stegosaurus* (right), drawn from models sold by the Natural History Museum, London. The plant is a cycad, such as *Triceratops* may have eaten. A large Black rhinoceros is also shown: it is 3.5 meters long, including the tail.

ently from the teeth of ornithopods. The upper and lower teeth did not meet crown-to-crown when the mouth closed, so they could not crush or grind food. Instead, the lower teeth came up close inside the upper ones (figure 1.12). We can be sure that they worked like this because the worn surfaces of the teeth are vertical. Thus the rows of teeth moved like the blades of shears. They would have been very effective for chopping up fibrous plant food, perhaps leaves of palms or cycads.

The pachycephalosaurs are a rather rare group of bipedal ornithischains. They have astonishingly thick skull roofs: some are 23 centimeters thick of solid bone. I have written about the function of this thickening in chapter 6.

The stegosaurids, another group of ornithischian dinosaurs, have a row of big bony plates along their backs (figure 1.13). I mention a possible function of these plates in chapter 7. Stegosaurids have long hind legs and short fore legs, but seem to have walked on all fours. Their heads are relatively small and their teeth are much less impressive than those of ornithopods and ceratopians. They have formidable spines on their tails.

The ankylosaurs are the last group of dinosaurs in this list. They had short legs and broad bodies, and walked on all fours. They had thick plates of bone embedded as armour plating in their skin, as if they were gigantic armadilloes. Some had club-like lumps of bone at the ends of their tails. I discuss them in chapter 5.

Dinosaurs lived on most of the earth's land mass throughout the Mesozoic era, but different dinosaurs lived in different places, at different times. *Tyrannosaurus* never tried its strength against *Apatosaurus*, which had been extinct for 70 million years before *Tyrannosaurus* evolved. *Compsognathus*, which lived in Europe, probably never met *Allosaurus*, which lived in the western United States, although they overlapped in time. Other groups of famous dinosaurs must have met. *Allosaurus*, *Apatosaurus* and *Stegosaurus* were all living in North America at the same time, late in the Jurassic period. Seventy million years later, late in the Cretaceous, *Tyrannosaurus, Anatosaurus,* and *Triceratops* were all living in North America.

Principal Sources

Charig (1979) is an excellent short introduction to the dinosaurs. Norman (1985) is a remarkable mine of information, but unfortunately includes no references to the more specialized scientific literature. Bakker (1986) is an idiosynchratic account of the dinosaurs by a scientist whose work has stimulated a great deal of interest and controversy. Czerkas and Olson (1987) shows how scientists' views about dinosaurs have changed in recent years. Norman and Weishampel (1985) describe the jaw mechanism of ornithopods. Paul (1988) compares the sizes of the largest dinosaurs. The other references in the list that follows are sources for illustrations.

Bakker, R. T. 1986. *The Dinosaur Heresies.* New York: Morrow.

Charig, A. 1979. *A New Look at the Dinosaurs.* London: British Museum (Natural History).

Coombs, W. P. 1980. Juvenile ceratopsians from Mongolia: The smallest known dinosaur specimens. *Nature* 283:380–381.

Czerkas, S. J. and E. C. Olson. 1987. *Dinosaurs Past and Present.* Seattle: University of Washington Press.

Galton, P. M. 1970. The posture of hadrosaurian dinosaurs. *Journal of Palaeontology* 44:464–473.

Newman, B. H. 1970. Stance and gait in the flesh-eating dinosaur *Tyrannosaurus. Biological Journal of the Linnean Society* 2:119–123.

Norman, D. B. 1985. *The Illustrated Encyclopedia of Dinosaurs.* London: Salamander Books.

Norman, D. B. and D. B. Weishampel. 1985. Ornithopod feeding mechanisms: Their bearing on the evolution of herbivory. *American Naturalist* 126:151–164.

Osborn, H. F. 1924. *Psittacosaurus* and *Protiguanodon*: Two Lower Cretaceous iguanodonts from Mongolia. *American Museum Novitates* 127:1–16.

Paul, G. S. 1988. The brachiosaur giants of the Morrison and Tendaguru. *Hunteria* 2(3):1–13.

II

Weighing Dinosaurs

THE DRAWINGS of dinosaurs in chapter 1 were based on fossil bones, fitted together to form complete skeletons. These skeletons tell us how large dinosaurs were. There is no reason to suspect that the sizes of bones have been changed by the processes of fossilization. There may of course be doubts about the size of the animal if bones are missing. For example, there are doubts about how many bones are missing from the tail of *Tyrannosaurus,* which one palaeontologist made 3.7 meters (12 ft) longer than another. There are also dangers of error if a complete skeleton has been assembled from bones of different individuals: animals of different sizes may be combined, producing an ill-proportioned skeleton. However, many of the best-known skeletons have been made from bones found together, apparently from one individual animal.

Measurements of skeletons tell us how long and how tall dinosaurs were, but do not directly tell us what they weighed. We might want to know, for at least two reasons. First, the weight (or mass) of an animal seems a fairer summary of its size than any measurement of height or length. Giraffes are taller than elephants and some pythons are longer but it is the elephant, the heaviest of the three, that impresses us as the largest. Second, some dinosaurs were so large as to make us wonder how easily their legs could have supported their weight. Were they as active as modern reptiles or were they lumbering monsters, barely able to move about? Were they perhaps so heavy that they had to live partly submerged in water, which would help to support them by buoyancy? To tackle questions like these, we want to know just how heavy the big dinosaurs were.

There is an awkward difference between the scientific meaning of the word "weight" and its use in everyday language. Non-scientists often give the "weights" of objects in pounds or kilograms, but scientists measure the quantity that they call weight in newtons. To us, weight means the force exerted on an object by gravity. It is measured in newtons, the unit of force. "Mass" is our name for the quantity that we measure in kilograms. (We don't use pounds.) Weight is calculated by multiplying mass by the gravitational acceleration *g* which is about 10 meters/second^2: that means that the speed of a falling body, unhindered by air resistance or anything else, increases by 10 meters per second every second. If a body has a mass of m kilograms, its weight on earth is $10m$ newtons.

Elephants, rhinoceroses, and (we will see) the larger dinosaurs have masses of several thousand kilograms. This makes it convenient to give their masses in tonnes (metric tons). The tonne is 1,000 kilograms or 2,205 pounds, almost the same as the commercial (long) ton of 2,240 pounds.

Modern animals can be weighed, but the masses of dinosaurs can only be estimated from other measurements. It seems sensible to give the masses of some large modern animals, as bases for comparison, before going on to the extinct ones.

Though it is possible to weigh modern animals, the biggest ones present problems. Domestic and zoo animals can be driven onto weighbridges like the ones used for weighing vehicles, but large wild animals are difficult to transport (alive or dead) to such facilities. Most available masses of large wild animals are of specimens shot and weighed in the field. Some were shot in the course of population control or scientific research, and others by hunters. A team culling a hippopotamus population used a two-tonne spring balance mounted on a hydraulic hoist on the back of a truck, but large animals usually have to be cut up and weighed in pieces. Some blood and other fluids are lost in the process, but the loss is generally no more than 3 percent of body mass.

Some masses of modern animals are given in table 2.1. All (except the humans) were wild animals killed in the field. Elephants continue growing throughout life and the masses given for them are the limits that they seem to approach in old age. The other masses are means from groups of adults. The Blue whale is the largest of the whales. The African elephant grows larger than the Indian elephant, which is not included in the table. The Black rhinoceros, however, is not the largest of the rhinoceroses. I have unfortunately not managed to find reliable masses for the White rhinoceros, which is said to surpass three tonnes. The lion is a large terrestrial carnivore, but not the largest. Tigers grow a little heavier, and the larger species of bear exceed half a tonne.

Now we return to the dinosaurs. How can we weigh them? In most cases, we have only their bones to guide us. We cannot estimate the masses of the living animals from the masses of their bones, because these have been altered by the processes of fossilization. They have lost water and protein and become impregnated with minerals.

We want to estimate not just the mass of the skeleton but its mass with all the guts, flesh, and skin that went with it. It seems easiest to base our calculations on models of the animals as we think they would have looked in life. Many such models have been made. The best commercially available ones that I know are plastic models at 1:40 scale, sold by the Natural History Museum, London. These seem to have been made carefully and accurately. I have checked many of their dimensions and find that they are indeed about 1:40 of the corresponding dimensions of the best-known fossil skeletons of the species they represent. I have used these models for estimating dinosaur masses. However, I once wanted to estimate the masses of moas (extinct ostrich-like birds from New Zealand) and could not find suitable models. I modeled the main features of the skeleton to scale in soldered wire, and used modeling clay to build up the flesh around it.

The masses of models depend on the densities of the materials used to make them, which may not be the same as the tissues of the animals. For this reason, the first stage in finding dinosaur masses is to measure the volume, not the mass, of the model.

Edwin Colbert, a distinguished U.S. paleontologist, measured the volumes of models in this way. He put a model in a box and packed sand around it until the box was filled to the brim. Then he removed the model, being careful to spill none of the sand, and added more sand until the box was full again. This extra sand had the same volume as the model.

TABLE 2.1. Masses (in tonnes) of some modern animals.

	Males	*Females*
Blue whale, *Balaenoptera musculus*	91	110
African elephant, *Loxodonta africana*	5.45	2.77
Hippopotamus, *Hippopotamus amphibius*	2.52	2.13
Black rhinoceros, *Diceros bicornis*	1.17	1.08
Eland, *Taurotragus oryx*	0.84	—
African buffalo, *Syncerus caffer*	0.75	—
Lion, *Panthera leo*	0.18	0.15
Human, *Homo sapiens*	0.07	0.05

SOURCES: whale: Nishiwaki 1950; elephant: Laws 1966; others: Meinertzhagen 1938.

Colbert's method was difficult to perform accurately because it depended on the sand being leveled off precisely at the top of the box. He used it in preference to the method that I used later, because he did not want to get his valuable plaster models wet. For measurements on plastic models, I used a method that depends on Archimedes' Principle, the principle of buoyancy. When an object is submerged in water, the water exerts an upward force on it. This force (called the upthrust) equals the weight of as much water as would have the same volume as the submerged body.

The diagram (figure 2.1) shows the method. The model is suspended by a thread from one arm of a beam balance, with a metal weight hanging from its tail. (The weight is unnecessary if the model is denser than water.) It hangs in a tall jar, not touching the bottom or the sides. Initially, the weight (if any) is submerged in water but the whole model is above the water. Enough weights are put on the pan to balance the system. Then water is added until the model is completely submerged. The upthrust of the water on the model puts the system out of balance and weights must be removed from the pan, to restore the balance. For example, when the experiment was done with a model of *Brachiosaurus*, weights totaling 728 grams had to be removed. This showed that the volume of the model equaled the volume of 728 grams water: it was 728 cubic centimeters.

The volume of the model must be scaled up to get the volume of the dinosaur. This particular model was 1:40 scale. Therefore the dinosaur was 40 times as long, 40 times as wide and 40 times as high as the model. Its volume was $40 \times 40 \times 40 = 64{,}000$ times the volume of the model: it was $64{,}000 \times 728 = 46{,}600{,}000$ cubic centimeters or 46.6 cubic meters.

To get the mass of the dinosaur from its volume, we must estimate its density. Most animals have about the same density as water. They either just float in water, with very little projecting above the surface, or just sink. Among living animals, crocodiles probably give the best indication of the probable density of dinosaurs. Not only are they among the largest modern reptiles, but they are believed to be quite closely related to the dinosaurs. Hugh Cott, a British zoologist, measured the densities of nine dead Nile crocodiles and got a mean value of 1,080 kilograms per cubic meter. This is quite a lot greater than the density of water, which is 1,000 kilograms per cubic meter. However, the lungs of the dead crocodiles were probably deflated, and live ones would have been less dense. Cott observed that crocodiles often float in water with only the nostrils and top of the head above the surface. They must then be very slightly less dense than water. He also observed that they sometimes rest motionless on the bottoms of rivers, when they must

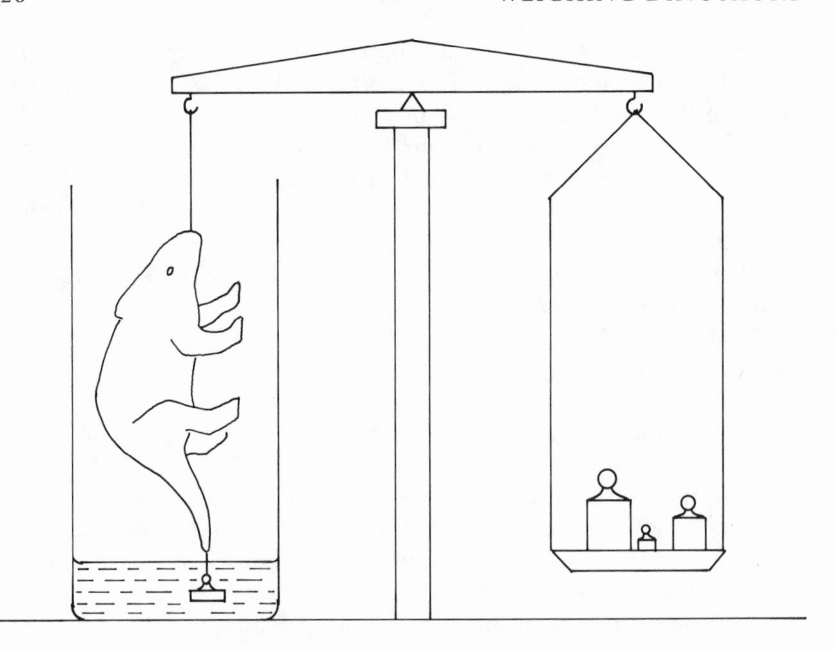

FIGURE 2.1. Apparatus for measuring the volumes of model dinosaurs.

be denser than water. Plainly, crocodiles can vary their density by inflating and deflating their lungs, but they are probably always very close to the density of water. I will therefore assume that dinosaurs had a density of 1,000 kilograms per cubic meter. Colbert assumed a smaller density, based on measurements on small reptiles that were probably inaccurate.

The volume of *Brachiosaurus* was estimated to be 46.6 cubic meters. If its density was 1,000 kilograms per cubic meter its mass was 46,600 kilograms, or 46.6 tonnes. This is colossal. It is about nine times the mass of a large male African elephant, the largest modern terrestrial animal. It is only half the mass of a large Blue whale, but whales live submerged in water which supports their weight by buoyancy.

The method of calculating masses from the volumes of models requires a fairly complete skeleton, unless the model maker is prepared to risk guessing the sizes of missing parts. An alternative approach using the dimensions of just a few selected bones was used in a recent international project. The collaborators were J. F. Anderson, a U.S. zoologist with a long-standing interest in the sizes of bones of different-sized modern animals; A. Hall-Martin, who works at Kruger National Park, South Africa; and Dale Russell, a Canadian dinosaur specialist.

Chapter 3 will show that dinosaurs stood and moved much more like mammals than like crocodiles, lizards, and tortoises. The Anderson team chose to study mammal bones to look for rules that would enable them to estimate dinosaur masses. The University of Florida, where Anderson works, has a large collection of mammal skeletons with records of the masses of the intact animals. That collection includes few really large mammals, but Hall-Martin was able to measure the bones of animals shot for other purposes in Kruger National Park.

They chose to study major leg bones, which are often well preserved in otherwise incomplete fossils. They could have used the lengths of the bones, but this might have led to errors due to some animals having spindly legs and others stumpy ones. An extreme case will show how serious this could be. Figure 2.2 shows two birds measured in Kenya by Professor Geoffrey Maloiy and me. The Secretary bird was slightly lighter than the Ground hornbill, but its leg bones were up to twice as long. The diameters and circumferences of leg bones seem to be better indicators of body mass, than are their lengths.

The Anderson team chose to study the circumferences of the humerus and femur, the bones of the upper parts of the legs (figure 2.3). They measured the circumferences of these bones where they were least, about half way along the shafts of the bones, and (in their studies of quadrupeds) added the two circumferences together. They could have used the circumference of either bone alone, but that would have led

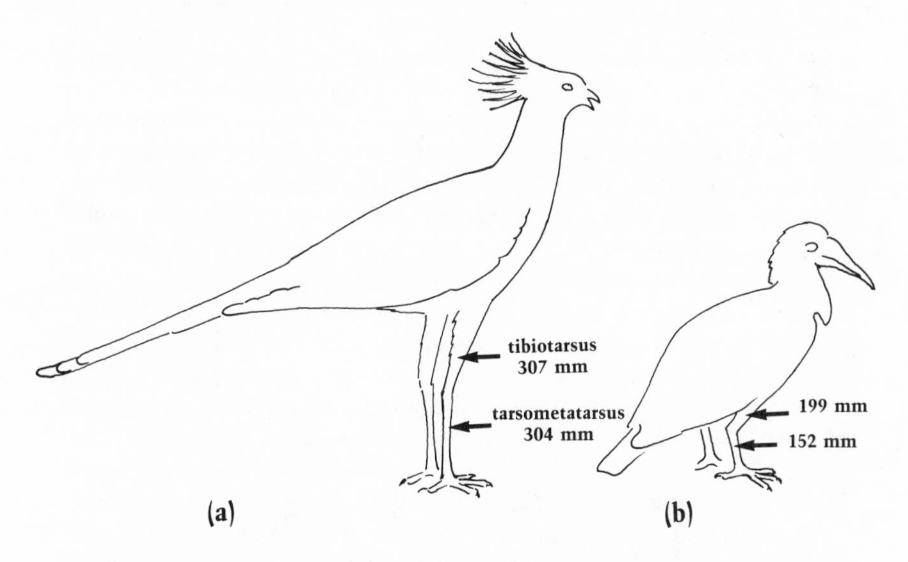

FIGURE 2.2. (a) a 3.9-kilogram Secretary bird (*Sagittarius serpentarius* and (b) a 4.2-kilogram Ground hornbill (*Bucorvus leadbeateri*). Their masses are about the same but the lengths of their leg bones are very different.

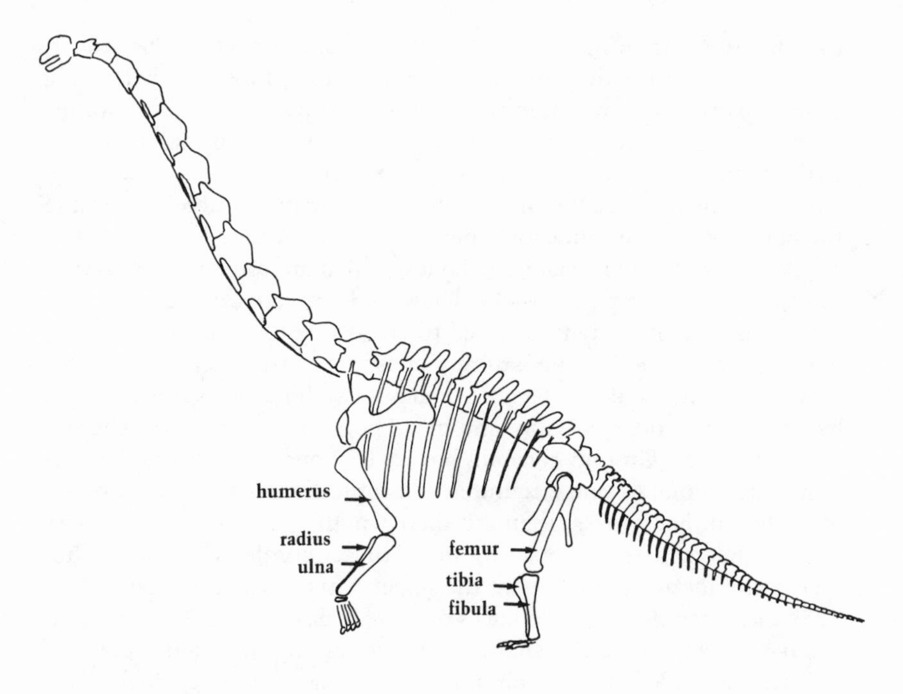

FIGURE 2.3. The skeleton of *Brachiosaurus brancai,* showing the names of some bones.

to errors due to different quadrupeds supporting different fractions of their weight on their fore and hind legs. *Apatosaurus* seems to have carried most of its weight on its hind legs, and has a thick femur and a thinner humerus. *Brachiosaurus* seems to have carried a bigger fraction of its weight on its fore legs, and the two bones are more nearly equal in circumference. Any formula based on one bone alone would probably overestimate the mass of one of these dinosaurs, and underestimate the other.

Figure 2.4 shows the total of humerus and femur circumferences plotted against body mass, for quadrupedal mammals. The scale of the graph has been chosen so that an exceptionally large elephant can just be fitted in, but that leaves most of the points for other animals squashed into the bottom left corner. It has been impossible to show the data clearly for anything smaller than a 29-kg baboon.

Figure 2.5 shows the same data (and more) plotted in a different way. The bottom scale has been distorted so that the distance from the 10-gram mark to the 100-gram mark is the same as from 100 grams to 1,000 grams (1 kilogram) or even from 10 tonnes to 100 tonnes. The scale up the side has been distorted in the same way. More precisely, distances along both scales are proportional to the logarithms of the numbers they represent, not to the numbers themselves. This has made it possible to show data for the whole range of sizes from mice to dinosaurs.

The solid black points in figure 2.5 are data for quadrupedal mammals. They form a more-or-less straight band. The line has been fitted to them by statistical analysis. It is the best straight line that can be got from this set of data, for estimating body mass from bone circumferences. Its equation (written in the most convenient form for the purpose) is

$$\text{body mass in kg} = 0.000084\ (\text{total of circumferences in mm})^{2.73}$$

(If you want to use this equation to estimate the body mass of a particular animal from its humerus and femur circumferences, you will have to use the y^x button on your calculator to raise the total circumference to the 2.73 power.)

All the black points lie close to the line, which suggests that it should be possible to estimate the masses of mammals rather accurately, from

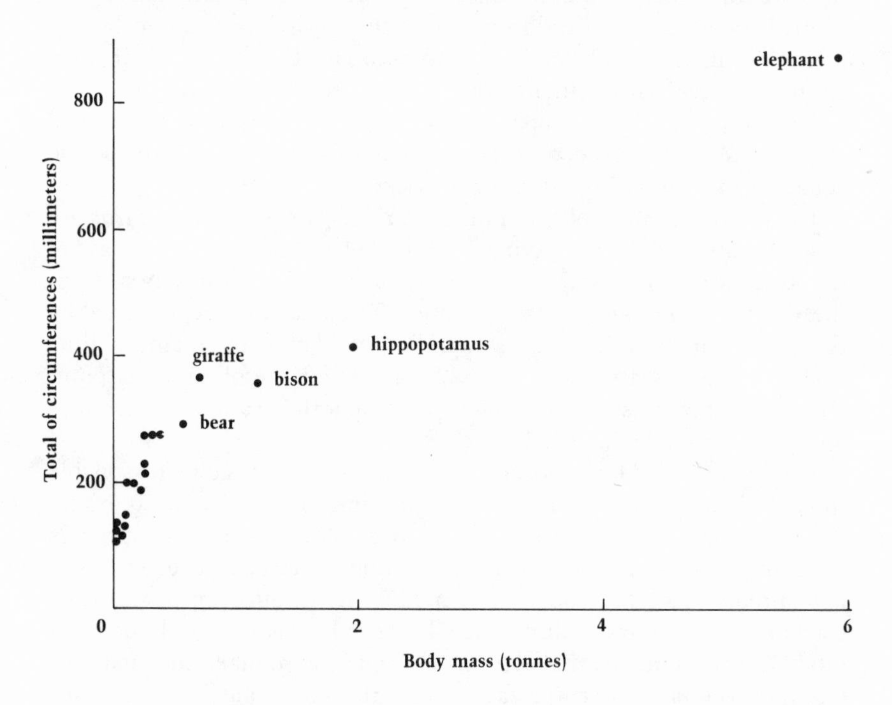

FIGURE 2.4. A graph showing (humerus circumference plus femur circumference) plotted against body mass, for various quadrupedal mammals. Data from Anderson et al. 1985.

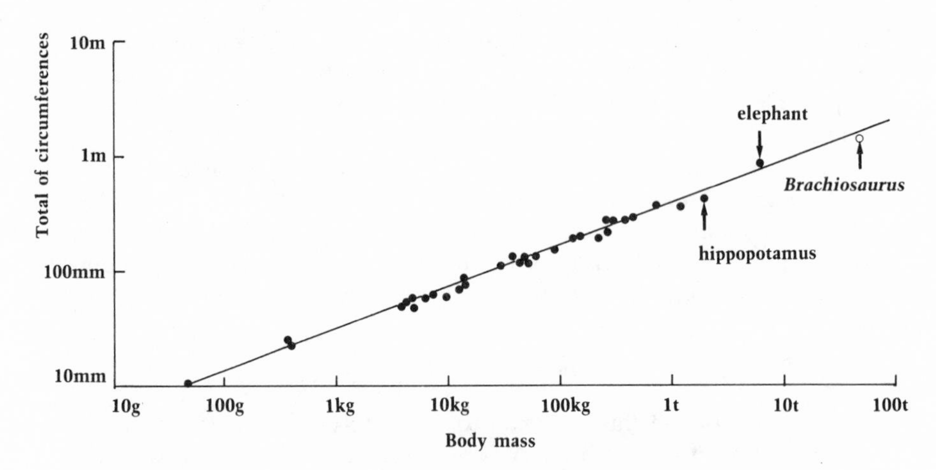

FIGURE 2.5. The same data as in figure 2.4 re-plotted on logarithmic scales, with additional points, including one for *Brachiosaurus*.

the circumferences of their bones. However, appearances can be deceptive. Imagine that we do not know the masses of the two largest mammals in the graph, and want to estimate them from their bone circumferences. The estimates that the equation would give us are 1.2 tonnes for the 2.0-tonne hippopotamus, and 9.0 tonnes for the 5.9-tonne elephant. We must expect errors as bad as this, or worse, if we use the equation for estimating dinosaur masses.

The circumferences of the humerus and femur of *Brachiosaurus* are 654 and 730 millimeters, giving a total 1384 millimeters. The mass of the same individual *Brachiosaurus* has already been estimated, from the volume of a model, to be 47 tonnes. These data are represented by the hollow circle in figure 2.5, which lies well below the mammal line. The line predicts a mass of only 32 tonnes, for the observed circumferences, but the discrepancy is no worse than the examples of the hippopotamus and elephant led us to expect.

The line is based on quadrupedal mammals, and it seems reasonable to apply it to quadrupedal dinosaurs. It would be a mistake to expect it to give accurate estimates of body mass, but even rough estimates are interesting. For bipeds, a different line is needed. The Anderson team produced a biped equation by modifying the quadruped one. They used only the femur circumference (because bipeds walk only on their hind legs) and adjusted the factor in the equation to make the predicted mass for one particular dinosaur match the mass that had been estimated from a model. Their biped equation is

$$\text{body mass in kg} = 0.00016\,(\text{femur circumference in mm})^{2.73}.$$

This underestimates the masses of kangaroos and overestimates the masses of ostriches.

Table 2.2 shows the masses of dinosaurs estimated from the volumes of models by Colbert and myself, and calculated from their equations by the Anderson group. For each of the species in the table, at least two estimates of mass have been made. For some of them, the estimates agree well, but for others there are big discrepancies. In the worst cases (*Diplodocus* and *Brachiosaurus*) the largest estimate is about three times the smallest.

Some of the discrepancies can be explained. Photographs in Colbert's paper show that his model of *Stegosaurus* was skinnier than the one I used, and his model of *Triceratops* was more portly than mine. In each case, one model may be more realistic than the other, but it is hard to say which. Even if a model is based on accurate skeleton measurements, its volume depends a lot on the judgement of the modeller. Some other discrepancies may be due to estimates being based on different-sized specimens of the same species.

TABLE 2.2. Masses (in tonnes) of dinosaurs.

	Colbert[a] *(1962)*	*Alexander*[a] *(1985)*	*Anderson et al.*[b] *(1985)*
theropods			
Allosaurus fragilis	2.3	—	1.4
Tyrannosaurus rex	7.7	7.4	4.5
sauropods			
Diplodocus carnegiei	11.7	18.5	5.8
Apatosaurus louisae	33.5	—	37.5
Brachiosaurus brancai	87.0	46.6	31.6
ornithopods			
Iguanodon bernissartensis	5.0	5.4	—
'Anatosaurus' copei	3.4	—	4.0
stegosaurs			
Stegosaurus ungulatus	2.0	3.1	—
ceratopians			
Styracosaurus albertensis	4.1	—	4.1
Triceratops 'prorsus'	9.4	6.1	—

NOTE: The Colbert estimates, as originally published, were based on an assumed density of 900 kilograms per cubic meter. They have been adjusted to a density of 1,000 kilograms per cubic meter, the same as for the Alexander values. The Anderson values for quadrupeds are also slightly larger than those given in the original paper, because an error in the equation has been corrected. The *Brachiosaurus* estimate in the Alexander column does not appear in Alexander (1985) but was obtained by the same method.

[a]Estimated from the volumes of models.

[b]Estimated from the circumference of bones.

It is hard to decide which of the mass estimates to prefer, in table 2.2. You might prefer the estimates based on models because they take account of the whole skeleton, not just the leg bones. Alternatively you might prefer the Anderson estimates because they do not depend on the judgment of a modeler. I am inclined to prefer the estimates from models.

Table 2.2. shows that there is uncertainty about the masses of dinosaurs. However, it is quite clear that the largest dinosaurs were exceedingly heavy. Compare this table with table 2.1, which gives the masses of modern animals. Whichever estimate you prefer, the bulky sauropods (*Apatosaurus* and *Brachiosaurus*) were many times as heavy as the biggest African elephants. The very long but slender *Diplodocus* was probably also much heavier than elephants: the sketch of one beside a cable car (figure 1.9) makes it hard to believe the very low Anderson estimate of its mass. *Triceratops,* the big ceratopian, seems to have been at least as heavy as most elephants. The biggest bipeds, both carnivores such as *Tyrannosaurus* and herbivores such as *Iguanodon,* were in the same range of masses, and many other dinosaurs were at least as heavy as an adult hippopotamus.

Finally, please remember that many dinosaurs were a great deal smaller than the ones in the table. The young *Psittacosaurus* in figure 1.2 is smaller than the pigeon.

Principal Sources

The estimates of dinosaur masses come from the papers of Colbert, 1962, Alexander 1985, and Anderson et al. 1985. The other papers in the list below are sources for the masses of modern animals given in table 2.1 and for figure 2.3.

Alexander, R. McN. 1985. Mechanics of posture and gait of some large dinosaurs. *Zoological Journal of the Linnean Society* 83:1–25.

Anderson, J. F., A. Hall-Martin, and D. A. Russell. 1985. Long bone circumference and weight in mammals, birds and dinosaurs. *Journal of Zoology* (A) 207:53–61.

Colbert, E. H. 1962. The weights of dinosaurs. *American Museum Novitates* 2076:1–16.

Janensch, W. 1950. Die Skelettrekonstruktion von *Brachiosaurus brancai. Palaeontographica* supplement 7:95–103.

Laws, R. M. 1966. Age criteria for the African elephant, *Loxodonta africana. East African Wildlife Journal* 4:1–37.

Meinertzhagen, R. 1938. Some weights and measurements of large mammals. *Proceedings of the Zoological Society of London* 1938A:433–439.

Nishiwaki, M. 1950. On the body weights of whales. *Scientific Reports on Whales of the Research Institute Tokyo* 4:184–209.

III

Dinosaur Footprints

OF ALL that remains of dinosaurs, their footprints bring the animals most vividly to life. Fossil bones may remind us of a rotting carcass, but footprints are evidence of a living, moving animal. It may seem incredible that dinosaur footprints should be preserved. The chances of an individual footprint surviving are remote, but millions of dinosaurs made footprints every day, for millions of years. For some, the improbable happened: they survived and were found by inquisitive people.

Figure 3.1 shows part of a remarkable group of over 3,000 footprints, found in a patch of rock in Australia. It seems obvious that they are footprints but they must be about 100 million years old. Geological evidence shows that the rock was formed in the middle part of the Cretaceous period, when dinosaurs were plentiful. The tracks of many individuals can be followed across the rock but none show any indication of distinct fore and hind feet, so the animals must have been bipeds. The biggest prints, 64 centimeters (25 inches) long, must have been made by a dinosaur: there were no other big bipeds around. The others, some of them as small as chicken footprints, were probably made by smaller dinosaurs. They are not quite the same shape as modern bird footprints, and bird fossils are rare in Cretaceous rocks.

The structure of the rock seems to show how the footprints got preserved. There are alternating layers of claystone (formed from mud) and sandstone (formed from sand). They seem to have been laid down where a river flowed into a lake, just as sand and mud carried down modern rivers get deposited as sand or mud banks at the mouth. In dry periods the lake would have shrunk leaving the mud bank exposed to the air. At times of flood the water would rise over the mud again and sediment would settle out of it. Coarse grains would settle first, forming sand, and fine grains would settle later, forming mud.

FIGURE 3.1. Dinosaur footprints found at Winton, Queensland, from Thulborn and Wade 1984.

A mud surface was exposed in a dry period when dinosaurs walked over it, forming footprints (figure 3.2a). The floods came and deposited a layer of sand, which filled the footprints (figure 3.2b). More layers of sediment were deposited, time passed and eventually the mud and sand became rock. The footprints had been impressed in mud and filled with sand, but now they were impressions in claystone, filled with sandstone. Fortunately the two kinds of rock do not stick together very firmly. Slabs of sandstone can be levered off leaving the claystone undamaged (figure 3.2c). After some of the prints had been found by a happy chance it was not too difficult to expose the rest. Some sandstone got left behind in the hollows of the prints, but it was possible to clean it out with an awl.

We have identified the tracks as those of bipedal dinosaurs, but we can be more specific. Both the flesh-eating theropods and the plant-eating ornithopods were bipedal, and both had three main toes. Theropods had sharp claws which were probably useful for attacking prey (figure 3.3a). Ornithopods had blunter, more hoof-like tips to their toes (figure 3.3b). The biggest footprints in figure 3.1 show claw marks, so were probably made by a theropod. Their size suggests an animal a little smaller than *Tyrannosaurus,* probably about five tonnes. Some of the smaller prints are thought to have been made by ornithopods and some by tiny theropods.

Figure 3.4 shows some footprints found in Texas. A few of them were made by three-toed bipeds, probably theropods, which were a little smaller than the big theropod at Winton. Most of the footprints, however, were made by quadrupeds. There are a lot of tracks on top of each other, giving a muddled impression, but one trackway (picked out in stipple) is very clear. It has huge hind footprints and smaller fore ones. Each is the right shape to have been made by a sauropod (figure 3.3c,d) and their size suggests an animal of 20–30 tonnes. The biggest of the hind footprints are 76 centimeters (30 inches) long. There are also footprints of smaller sauropods.

Even bigger footprints have been found elsewhere in Texas, with hind prints 92 centimeters (36 inches) long.

The footprints in figures 3.1 and 3.4 are at exceptional sites, where large numbers of dinosaur footprints have been found together. Less impressive finds are quite common. Large numbers of dinosaur footprints have been found, all over the world. They can tell us a lot about the lives of dinosaurs.

First, how did dinosaurs stand and move? Modern reptiles walk with their feet well out on either side of the body, so the lines of footprints made by their left and right feet are well apart (figure 3.5a). Birds and mammals walk with their feet under the body, and the lines of left and right prints are close together (figure 3.5b,c). Dinosaur footprints show that they walked like birds and mammals, with their feet well under the body. The shapes of dinosaur leg bones show the same thing. They would have dislocated their joints if they had tried to walk like modern reptiles.

The footprints in figure 3.6 show an exception to the general rule. There are large, ill-formed hind footprints (stippled) with smaller fore prints (black) on either side. The prints are too poor to show the shapes of the feet that made them, but they were probably made by *Iguanodon,* which was around at the time. The hind feet were apparently under the body, mammal-fashion, but the fore feet reached out to either side. *Iguanodon* looks like a biped, but these tracks suggest that its fore feet took a little of its weight.

One striking thing about dinosaur footprints is that there is seldom any sign of marks made by the tail. Old pictures of dinosaurs show the

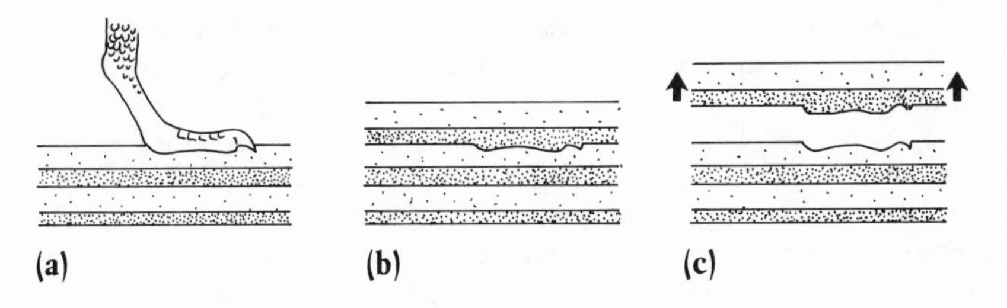

FIGURE 3.2. Diagrams showing how the Winton footprints were preserved. Each diagram is a vertical section through the layers of mud (light stipple) and sand (dense stipple).

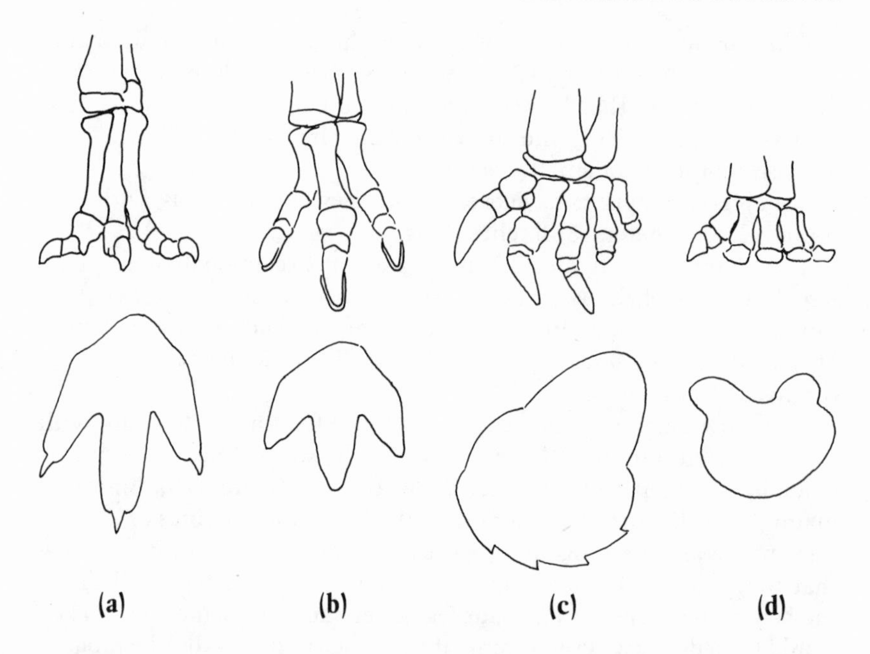

FIGURE 3.3. Skeletons of dinosaur feet: (a) *Tyrannosaurus* hind foot; (b) *Iguanodon* hind foot; (c), (d) *Apatosaurus* hind and fore feet. The outline under each skeleton shows the probable shape of its footprint.

tails of most of them dragging on the ground. There is more about this in chapter 5.

Some sauropod footprints are 10–15 centimeters (4–6 inches) deep. They were probably deeper when first formed because the mud must have lost water as it compacted to form rock, but they cannot have been very much deeper. If there had been too much water in the mud, it would have been too sloppy to take a footprint. The ground must have been fairly soft, but not sloppy.

Soft ground may have been dangerous for big dinosaurs, because of their size. Imagine two dinosaurs of identical shape, one twice as long as the other. It is twice as long, twice as wide and twice as high, so it is eight times as heavy ($2^3 = 8$). The soles of its feet are twice as long and twice as wide, so they have four times the area ($2^2 = 4$). Eight times the weight has to be supported on four times the area, so the pressure under the big dinosaur's feet is twice as much as under the small one. The big dinosaur is more likely to sink in and get bogged down.

The argument seems clear, but it is a bit too simple. Suppose there were a thick layer of soft mud with firm ground below. Small animals might get bogged down but big ones with longer legs might be able to walk, with their feet sinking to the firm ground.

There is another complication. Different soils support loads in different ways. Wet clays depend mainly on cohesion between the clay particles, and the load they can support is about proportional to the area carrying the load. Dry sand depends mainly on friction between the grains and the load it can support is about proportional to the 1.5 power of the area: that means that four times the area can support eight times the load. Our big dinosaur would be more likely than the small one to get bogged down in wet clay, but the danger of sinking in dry sand would be about the same for both.

Real animals of different sizes are not the same shape, like these imaginary dinosaurs. Table 3.1 shows masses and foot areas of various animals, based on the best data I can find. Here is how the table works. The mass of a particular *Apatosaurus* was probably about 35 tonnes or 35,000 kilograms (table 2.2). Its weight (mass multiplied by gravitational acceleration) was therefore 35,000 × 10 = 350,000 newtons or 350 kilonewtons. Some of the biggest known sauropod footprints are about the right size to have been made by it. The area of each fore footprint is 0.16 square meters and that of each hind print is 0.43 square meters, giving a total (two fore and two hind feet) of about 1.2 square meters. When the animal stood, this area supported 350 kilonewtons, so (weight/area) was 290 kilonewtons per square meter. It is easy to

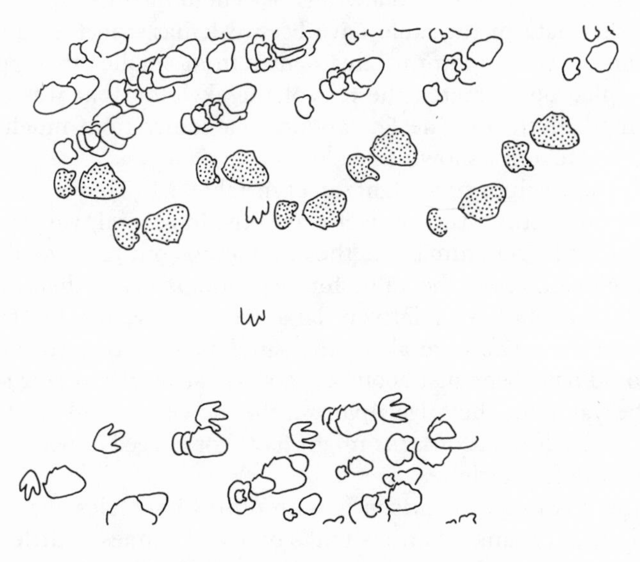

FIGURE 3.4. Footprints at Davenport Ranch, Texas, redrawn from Bird (1944). The rock was formed in the early part of the Cretaceous period.

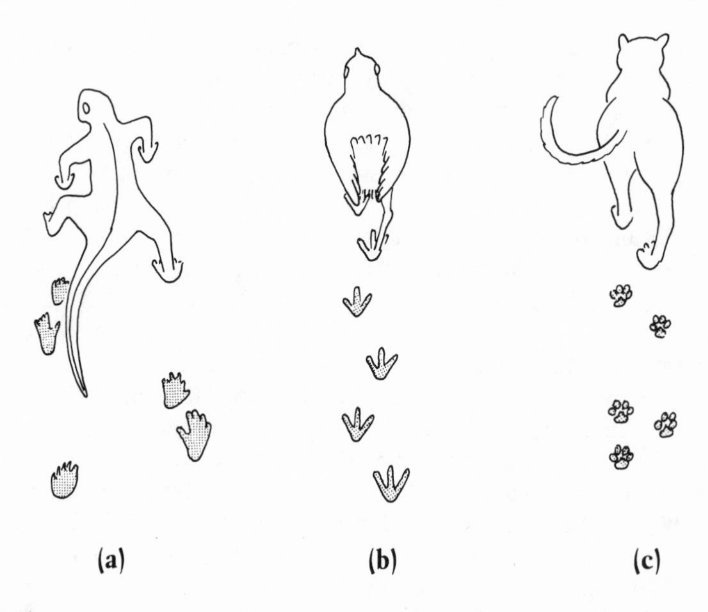

FIGURE 3.5. Rear views of (a) a lizard; (b) a bird; and (c) a mammal, and their footprints.

calculate that $1.2^{1.5}$ is 1.31 (use the y^x button on your calculator) so weight/$(\text{area})^{1.5}$ was 270 kilonewtons per cubic meter.

The other data in the table have been calculated in the same way. The *Tyrannosaurus* footprint area comes from smaller theropod footprints, scaled up to match the feet of the skeleton that was used for estimating body mass. The *Iguanodon* area comes from much clearer prints than the ones shown in figure 3.6.

Look at the values of (weight/area) in table 3.1. They tell us about the danger of getting stuck in wet clay soils. In general, we expect bigger values for bigger animals, as the argument about the two dinosaurs showed. Nevertheless, the value for elephants is lower than for cattle because elephants have relatively large feet. The values for *Tyrannosaurus* and *Iguanodon* are about the same as for cattle, so these animals would have been just about as good as cattle, at crossing soft wet clay. The value for the huge *Apatosaurus*, however, is about twice as high as for cattle. *Apatosaurus* might have got bogged down on ground that was safe for cattle.

Another possible comparison is with off-road vehicles such as tractors and military tanks. Various tanks of 37–51 tonnes (a little heavier than *Apatosaurus*) have peak pressures of 200–270 kilonewtons per square meter, under their tracks. These seem close to *Apatosaurus'* value of 290 kilonewtons per square meter until you realize that *Apatosaurus* would have to lift its feet in turn when it walked. Peak forces on the feet of people and animals during walking are generally about double the standing values. The pressure under the feet of *Apatosaurus* must have reached 580 kilonewtons per square meter, when it walked. It seems that the dinosaur would be less good than a tank at crossing soft ground.

Now look at the values for weight/(area)$^{1.5}$, which tell us about the danger of sinking in dry sand. In this case we have to be careful about our comparisons. Equal values mean equal danger of sinking only if the animals have feet of about the same shape, and the same number of feet. *Apatosaurus*, elephants and cattle all have four, roughly circular, feet, so comparisons between them seem fair. *Apatosaurus* has a higher value than elephants but a much lower one than cattle. It would be much less likely than a cow to get stuck in a sand dune.

Most real soils have properties between the extremes of wet clay and dry sand.

Fossil footprints can also tell us about the speeds of the dinosaurs that made them. They cannot tell us as certainly as if we could watch dinosaurs running and time them with stopwatches, but they can give us estimates that are probably fairly reliable.

When people walk slowly they take short strides. When they walk faster they take longer strides and when they run they take still longer

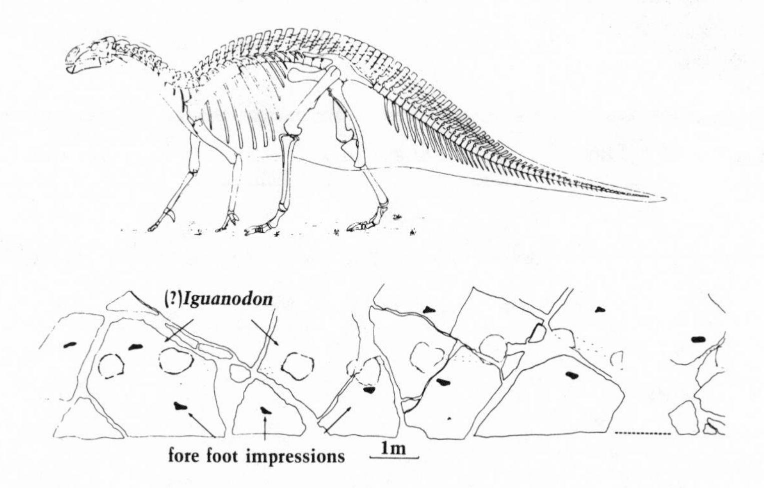

FIGURE 3.6. (above) *Iguanodon bernissartensis* walking with its fore as well as its hind feet touching the ground and (below) footprints which it may have made. From Norman 1980. Both the skeleton and the footprints are from early Cretaceous rocks.

strides (figure 3.7). Notice how stride length is defined: it is the distance from one footprint to the same point on the next print of the same foot. Figure 3.8 shows stride lengths and speeds for human adults. If you find a set of footprints you can measure the stride length and use this graph to estimate how fast the person was going.

The same graph shows the same thing for some animals. The faster they go, the longer their strides. Watch an adult walking with a small child. The adult takes a few long strides while the child takes a lot of short ones. Similarly, small animals take shorter strides than large ones, at the same speed. Dogs take shorter strides than camels (figure 3.7).

Does this mean that to estimate speeds from stride lengths we need separate graphs for each species, and separate graphs for adults and young of each species? If that were true, dinosaur footprints could tell us nothing about dinosaur speeds because we could never get the data for the special graphs for dinosaurs. Fortunately, it seems not to be true. There is a general rule that applies to birds and mammals and probably also to dinosaurs. It seems better to compare dinosaurs to birds and mammals than to modern reptiles, because dinosaur footprints show that they walked with their feet well in under the body (figure 3.5).

How then should we look for a general rule? What we want to do is to make a graph like figure 3.8 that will apply to animals of different sizes. We expect long-legged animals to take long strides and short-legged animals to take short ones, so it seems sensible to calculate *relative* stride lengths

relative stride length = (stride length)/(leg length).

TABLE 3.1 Data about the danger of getting stuck in soft ground.

	MASS	FOOT AREA	$\frac{\text{WEIGHT}}{\text{AREA}}$	$\frac{\text{WEIGHT}}{(\text{AREA})^{1.5}}$
	Tonnes	*Square Meters*	kN/m^2	KN/m^3
Apatosaurus	35	1.2	290	270
Tyrannosaurus	7	0.6	120	150
Iguanodon	5	0.4	120	190
African elephant	4.5	0.6	70	90
Domestic cattle	0.6	0.04	150	740
Human	0.07	0.035	20	110

SOURCE: Data are from table 2.1 and Alexander 1985.

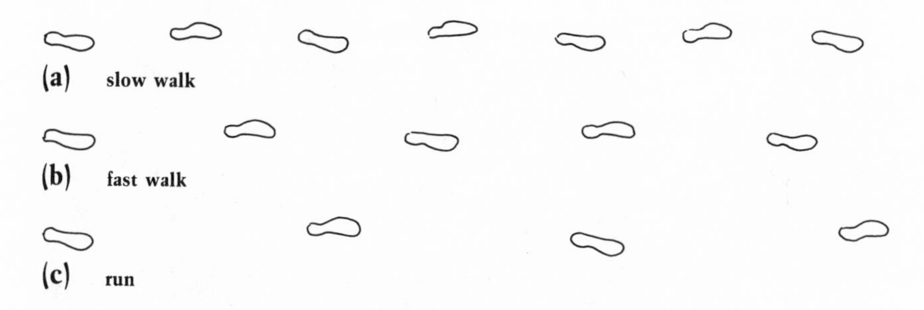

FIGURE 3.7. Footprints of a man walking slowly, walking fast, and running.

Leg length could be defined in various ways, but we will use the height of the hip joint from the ground in normal standing (figure 3.9).

We may expect animals of different sizes to have equal relative stride lengths, when running at equivalent speeds. That begs a question: what do we mean by equivalent speeds?

The same question faces ship designers, who want to do tests on models before building real ships. (That way they avoid expensive mistakes.) They want to build ships that are cheap to run, needing little power to propel them. Much of the power for driving a ship is needed to push along the bow wave, that builds up in front of the bow and spreads out to either side. When a model is being tested it must be moved at the right speed to make a bow wave of the same height (relative to the size of the model) as the wave that the real ship will make. This is not the same as the speed of the full-sized ship: it is an equivalent speed.

Physical theory says that equivalent speeds, for this purpose, mean equal dimensionless speeds, calculated in this way:

$$\text{dimensionless speed} = \frac{\text{speed}}{\sqrt{\text{hull length} \times \text{gravitational acceleration.}}}$$

Suppose a real ship, 300 meters long, is to travel at 15 meters per second. The gravitational acceleration is 10 meters/second^2 so the dimensionless speed is $15/\sqrt{300 \times 10} = 0.27$. If you want to make tests on a 5-meter model you should run it at 1.9 meters per second to get the same dimensionless speed: $1.9/\sqrt{5 \times 10} = 0.27$.

If that were a special rule that applied only to ships it would be no use to us. It is actually a special case of a general rule that applies to any motion in which gravity is important. Gravity is important in the motion of ships because it pulls downward on the bow wave, trying to flatten it. It is important in walking and running because of its effect on the rise and fall of the body.

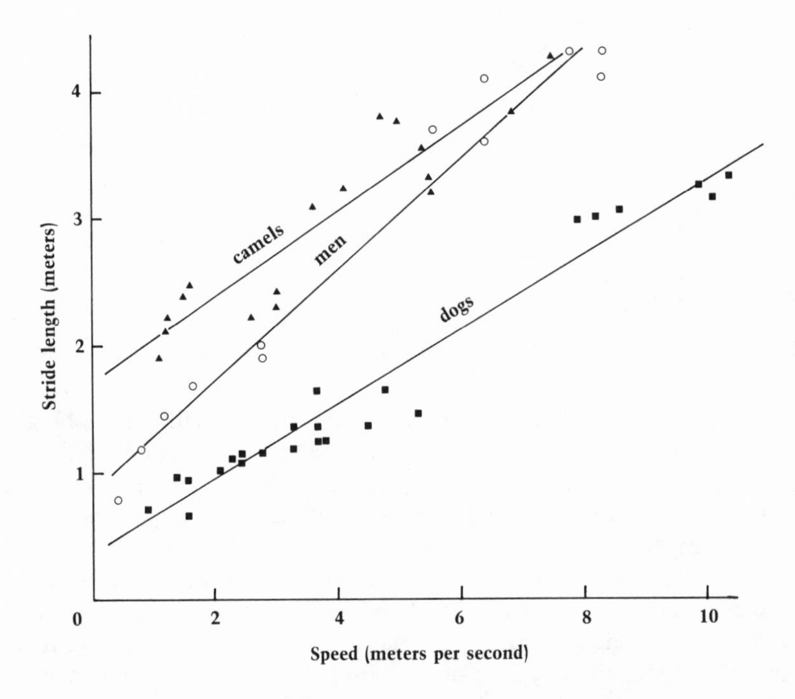

FIGURE 3.8. Graphs of stride length against speed for adult men: (○), dogs (■), and camels (▲).

The general rule uses a concept called dynamic similarity. Imagine you have films of two animals running, a large animal and a small one. Show these films side by side, using two projectors. You could adjust the sizes of the images, making the animals look the same size, by setting up the projectors at different distances from the screen. You could adjust the tempo of the animals' running, so that both seemed to take the same number of strides per second, by running the projectors at different speeds. You might find that the adjustments made the two films look identical. That would happen if the two animals moved in dynamically similar fashion. Dynamic similarity means that the movements of one could be made identical to those of the other by multiplying all lengths by one factor, all times by another factor, and all forces by a third factor.

Physical theory says that if gravity is important, dynamic similarity is only possible for animals (or ships, or anything else) moving with equal dimensionless speeds. Any suitable length can be used for calculating dimensionless speeds. Hull length is used for ships, but leg length is used for animals.

$$\text{dimensionless speed} = \frac{\text{speed}}{\sqrt{\text{leg length} \times \text{gravitational acceleration}}}$$

Animals moving with equal dimensionless speeds do not *have* to make dynamically similar movements, but there is a good reason for expecting them to. If they are to move in the most economical way possible, so that their muscles have to do least work, they must move in dynamically similar fashion. Among other things, this means that they must use equal relative stride lengths. If we plot graphs of relative stride length against dimensionless speed, we can expect to get near-identical graphs for similar animals of different sizes.

Figure 3.10 shows that we do. It includes all the data from figure 3.8 plus some more. It includes mammals ranging in size from dogs to an elephant, and even a bird (the ostrich). It includes bipeds (humans and the ostrich) as well as quadrupeds. Nevertheless, all the points lie reasonably near the line. Strictly speaking, dynamic similarity is only possible between animals of the same shape. Obviously, ostriches are not the same shape as people, dogs, or elephants, but all of these use about the same relative stride length, at any particular dimensionless speed. It seems likely that the same rule will also hold for dinosaurs. We can use the graph to get rough estimates of dinosaur speeds.

Before using it, we must get rid of one more worry. All the data are for animals running on hard ground, where they did not leave visible footprints. The dinosaur stride lengths that we know come from footprints left in ground that must have been fairly soft. Does the relationship between stride length and speed that holds on hard ground hold also on soft?

When I thought of this worry I was spending Christmas with my family in Norfolk, England. I took my children (then aged 11 and 13)

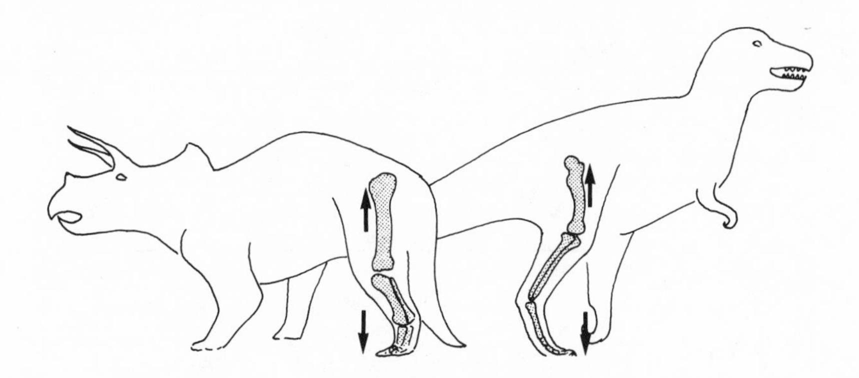

FIGURE 3.9. Diagrams showing the meaning of leg length.

to a nearby beach at low tide. The top of the beach was firm sand, and we left only slight footprints there. Half way down the beach was softer sandy mud, and we made fairly deep footprints. At the bottom, near the water's edge, was really horrible soft mud and we sank in almost to the tops of our rubber boots. At each of these levels we marked out a 25-meter course. We walked and ran over these courses at various speeds, in random order. We timed each other with stop-watches and counted our strides, so we were able to calculate stride lengths and speeds. We found that we could not run fast in the soft mud—indeed, we could hardly run in it at all—but at any speed we could manage, softness seemed to have no effect on stride length. Each of us used the same stride length at any particular speed, however firm or soft the ground. This encouraged me to hope that a graph like figure 3.10 would give reliable estimates of dinosaur speeds, even if the dinosaurs were on soft ground.

One more thing was needed, before the graph could be used. I needed a way of estimating dinosaur leg lengths from their footprints. To find one, I measured a lot of dinosaur skeletons. I measured the length of

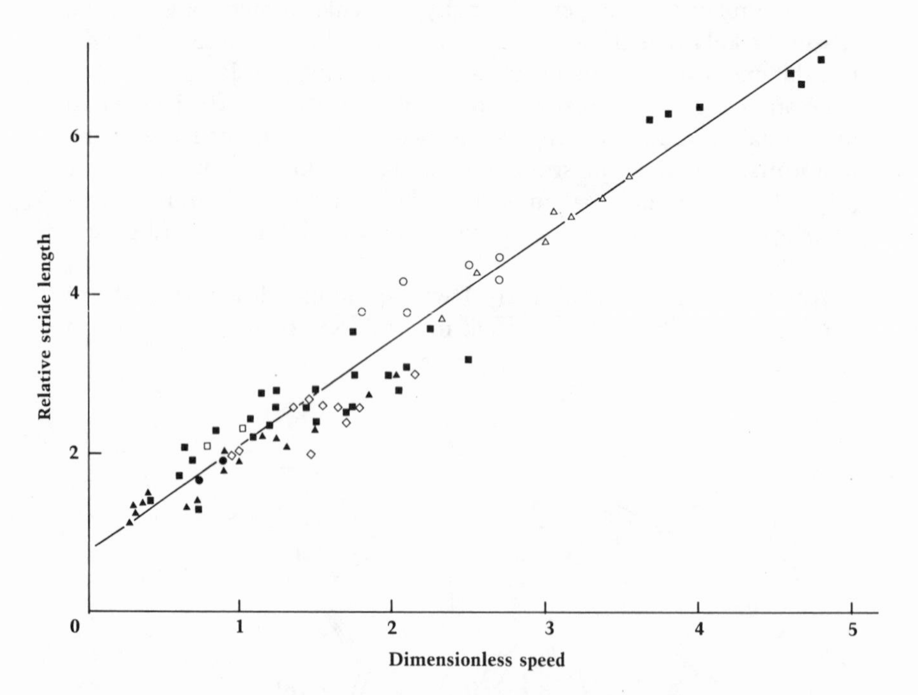

FIGURE 3.10. A graph of relative stride length against dimensionless speed for ostrich: (●), humans (○), dogs (■), elephant (□), rhinoceros (◇), sheep (△), and camels (▲). Data from Alexander 1976 and Alexander and Jayes 1983.

the part of the foot that would rest on the ground and make a footprint, and also the height of the hip joint from the ground. (There is a danger of error here, if the skeleton has been put together with the knee unnaturally bent or extended.) I found that leg length was about four times foot length in a wide variety of dinosaurs, both bipeds and quadrupeds, and decided to assume that as a general rule.

At last we are ready to estimate a speed. Take the case of the large theropod in figure 3.1. Its foot length (measured from the most complete prints) was 0.64 meter, so its leg length can be estimated as 2.56 meters. Its stride length was 3.31 meters so its relative stride length was 3.31/2.56 = 1.3. The graph (figure 3.10) shows that we can expect this relative stride length when the dimensionless speed is 0.4. For an animal of leg length 2.56 meters, that means a speed of 2.0 meters per second:

$$2.0/\sqrt{2.56 \times 10} = 0.4.$$

That estimate of speed is inevitably a rough one. The scatter of points around the line in figure 3.10 shows that we cannot expect accurate answers. Nevertheless it seems clear that the big theropod was traveling quite slowly. For humans, with much shorter legs, 2.0 meters per second (4.5 miles per hour) is only a brisk walking speed.

Some other speed estimates are shown in table 3.2. They have been obtained in the same way, except that the leg length estimates for the small Winton dinosaurs are not exactly four times footprint length: the ratio has been estimated for each group of dinosaurs. Also, some of the speeds have been calculated by a different formula that I published before I had collected the data for high speeds in figure 3.10. This makes little difference to the results.

Table 3.2 also shows whether each dinosaur seems to have been walking or running. People and many birds walk to go slowly and run to go faster. Similarly, horses, dogs, and other quadrupedal mammals walk at low speeds and use various running gaits (trotting, galloping, etc.) at higher speeds. There are various differences between walking and running but the most obvious is that when we walk we have each foot on the ground for more than half the time so there are stages in each stride when *both* feet are on the ground. In running, each foot is on the ground for less than half the time so there are stages when *neither* foot is on the ground. Similarly for quadrupeds walking, each foot is on the ground more than half the time so there are stages when both fore feet, or both hind feet, are on the ground. For quadrupeds running, each foot is on the ground for less than half the time so there are stages when neither fore foot, or neither hind foot, is on the ground.

Watch an animal speeding up, changing from a walk to a run. In most cases, the change is abrupt. A person or a horse, for example, is obviously walking at one moment and obviously running at the next. Some animals (for example, sheep) make the change more gradually, but whether the change is abrupt or gradual, it occurs at a highly predictable speed. People, and mammals in general, and birds walk when their dimensionless speed is less than about 0.7 and run when it is more than about 0.7. That rule has been used in table 3.2 to distinguish between walking and running. If the dimensionless speed estimated from the stride length is more than 0.7, the dinosaur is marked as running. It would be hard to be sure which the gait had been, if the estimated dimensionless speed were close to 0.7, but all the cases in the table seem clear cut.

Very few footprints of running dinosaurs are known, apart from the ones at Winton. This is not surprising. People and other fairly large mammals walk a lot, but run only occasionally. They are particularly unlikely to run on ground soft enough to take deep footprints, except in emergencies.

Though running dinosaur footprints are unusual, there are some that seem to show faster running. The fastest recorded are some theropod tracks in Texas: footprints of two dinosaurs, one with 29 centimeter feet and the other with 38 centimeter feet, indicate speeds of 12 meters

TABLE 3.2. Estimated speeds of dinosaurs.

	ESTIMATED LEG LENGTH	ESTIMATED SPEED		GAIT
	Meters	*Meters Per Second*	*Miles Per Hour*	
data from Davenport Ranch (fig. 3.4)				
large theropod	2.0	2.2	(4.9)	walk
small theropod	1.0	3.6	(8.1)	run
large sauropod	3.0	1.0	(2.2)	walk
small sauropod	1.5	1.1	(2.5)	walk
data from Winton (fig. 3.1)				
large theropod	2.6	2.0	(4.5)	walk
small theropods	0.13–0.22	3.0–3.5	(6.7–7.8)	run
ornithopods	0.14–1.6	4.3–4.8	(9.6–10.7)	run

per second (27 miles per hour). The larger of these, made by an animal of (probably) about 0.6 tonne, seems also to be the largest set of dinosaur footprints to show running. All footprints that I know of larger dinosaurs seem to show walking.

The top dinosaur speed, of 12 meters per second, is fast. Good human athletes sprint at up to 10 meters per second, racehorses at up to 17 meters per second, and greyhounds at up to 16 meters per second. These speeds are known accurately, because they are measured in races. Remarkably little is known about top speeds of other animals. With colleagues, I have measured the speeds of various African animals by filming them from a vehicle, while chasing them across country on the grassy plains of Kenya. Zebras, giraffes, and various antelopes galloped at top speeds of 11 to 14 meters per second. Ostriches seemed to be a little faster. It seems possible that the racers bred by man, racehorses and greyhounds, are the fastest land animals. This would not be surprising. They have been bred for speed alone, but the evolution of wild animals has been guided by other needs, as well as speed.

Many books give higher top speeds for wild animals, up to an amazing (incredible?) maximum of 30 meters per second for cheetahs. None of the books that give these speeds explain in detail how they were measured. There are many possible sources of error. For example, imagine you are driving in a vehicle alongside an animal, watching your speedometer. It swerves away from you, and you swerve to keep up with it. You are now traveling on the outside of a bend, with it on the inside. You must drive faster than it is running, to keep beside it. Errors of this kind were avoided in the experiments that gave top speeds of 11 to 14 meters per second. The methods used were explained in detail in a scientific journal, so that other scientists could criticise them if they were faulty.

That means that the estimated top speed of 12 meters per second, for the Texas theropods, would probably be a very creditable speed for an antelope. Those dinosaurs were not particularly large, but they were fast.

We will ask one more question of the footprints. The Winton footprints (figure 3.1) and the Davenport Ranch ones (figure 3.4) were made, in each case, by many individual dinosaurs. Were these dinosaurs moving around in groups and, if so, what were they doing?

In both cases they are suspected of moving in groups, but the evidence is inconclusive. In both cases there are a lot of tracks running in the same direction, which suggests group movement. All the prints seem to have been fresh when silted over and therefore (probably) were made about the same time. There is no indication of some older prints being more weathered by rain or other causes than later prints.

At Davenport Ranch there are tracks of 23 sauropods, all walking in the same direction. They all had similar-shaped feet and probably belonged to the same species, but some had feet twice as large as others. The largest were probably eight times as heavy as the smallest but all the sauropods, large or small, seem to have been walking at about the same low speed. The impression is of a herd, adults and juveniles, traveling together at a leisurely pace.

At Winton the picture seems more exciting. The single large theropod seems to have been walking southwest. More than 150 small dinosaurs, theropods and ornithopods, ran much faster in the opposite direction, toward the northeast. Richard Thulborn and Mary Wade, who studied the tracks, tell this story. The area was a mud bank at the edge of a lake. A large group of small dinosaurs were on the bank where they had come to drink or to hunt for food. The ornithopods may have been feeding on plants and the small theropods may have been traveling with them, ready to catch and eat any insects that they disturbed. Both the ornithopods and the theropods varied in size, but some of them were probably juveniles. The small theropods may all have been one species and all but one of the ornithopods may have been another species. (The exception had *much* larger feet than any of the other ornithopods.)

Suddenly the big theropod arrived, walking off the mainland onto the mudbank. It was much bigger than any of the other animals, well able to attack and eat any of them. The small dinosaurs were threatened, but the way forward was blocked by the water. All they could do was to double back past the theropod onto the mainland, running as fast as they could on the soft ground.

This chapter has told about fossil footprints of dinosaurs and has shown how the speeds of animals that made them have been estimated, from the spacing of the prints. Most of the speeds indicated by dinosaur footprints (especially by big ones) are slow, but a few footprints of running dinosaurs are known. Two exceptional trackways seem to have been made by dinosaurs running at speeds that would be fast for antelopes.

I also discussed whether big dinosaurs would be in danger of sinking in soft ground. *Apatosaurus* would have been more apt to sink than a cow in soft clay, but would be less likely than a cow to sink in a sand dune.

Principal Sources

The Davenport Ranch and Winton tracks have been described by Bird (1944) and Thulborn and Wade (1984), respectively. Norman (1980) discussed the supposed *Iguanodon* track. Farlow (1981) described the very fast theropod track.

Alexander (1985) collected the data about pressures on the ground. Alexander (1976) introduced the method of estimating speeds from stride lengths. The "Froude number" defined in that paper is the square of the quantity called "dimensionless speed" in this book. The data on stride lengths of modern animals comes largely from Alexander and Jayes (1983) and the data on maximum speeds from Alexander, Langman, and Jayes (1977).

Alexander, R. McN. 1976. Estimates of speeds and dinosaurs. *Nature* 261:129–30.

Alexander, R. McN. 1985. Mechanics of posture and gait of some large dinosaurs. *Zoological Journal of the Linnean Society* 83:1–25.

Alexander, R. McN. and A. S. Jayes. 1983. A dynamic similarity hypothesis for the gaits of quadrupedal mammals. *Journal of Zoology* 201:135–152.

Alexander, R. McN., V. A. Langman, and A. S. Jayes. 1977. Fast locomotion of some African ungulates. *Journal of Zoology* 183:291–300.

Bird, R. T. 1944. Did brontosaurus ever walk on land? *Natural History, New York* 53:60–67.

Farlow, J. O. 1981. Estimates of dinosaur speeds from a new trackway site in Texas. *Nature* 294:747–748.

Norman, D. B. 1980. On the ornithischian dinosaur *Iguanodon bernissartensis* of Bernissart (Belgium). *Mémoires de l'Institut Royal des Sciences Naturelles de Belgique* 178:1–103.

Thulborn, R. A. and M. Wade. 1984. Dinosaur trackways in the Winton formation (mid-Cretaceous) of Queensland. *Memoirs of the Queensland Museum* 21:413–517.

IV

Dinosaur Strengths

FOSSIL FOOTPRINTS seem to tell us that large dinosaurs usually moved slowly, but could they move fast in an emergency? Could they jump, and do other athletic things? We expect very large animals to be lumbering monsters. Buffaloes are less nimble than gazelles and elephants are less nimble than buffaloes. Were dinosaurs that were much larger than elephants less nimble still?

You have to be strong to run fast or to jump. When you stand, your weight is divided between your two feet: the force on each foot equals half your weight. When you walk, for much of the time you have only one foot on the ground and the peak force on each foot is about equal to body weight. When you run, each foot is on the ground for less time so the peak forces have to be bigger. The *average* vertical force, over a complete stride, must equal body weight, so if the feet are on the ground for less of the time the peak forces must be bigger. In jogging, each foot is on the ground for about 35 percent of the time and the peak force is about 2.7 times body weight. In sprinting, each is on the ground for about 22 percent of the time and the peak force is about 3.5 times body weight. These forces have been measured by means of force plates, force-sensitive panels set into the ground. The faster you run, the bigger are the forces, and even bigger forces may act when you jump.

Imagine two animals of exactly the same shape, one twice as long as the other. It is twice as long, twice as wide, and twice as high, so it is eight times as heavy. Its bones and muscles have twice the diameter, and four times the cross-sectional area. Suppose that these two animals perform the same activity, moving in dynamically similar fashion. All the forces involved are in proportion to their weight, eight times as large for the larger animal. The strengths of their bones and muscles are proportional to their cross-sectional areas, only four times as large for the larger animal. This means that the big animal is working nearer the limit than the small one, the limit set by the strengths of its bones and muscles. If the two animals try something more strenuous, the small animal may succeed but the large one may not.

That argument does not apply directly to real animals, because animals of different sizes are not the same shape, but it helps us to understand why buffaloes are less athletic than gazelles. It also makes it clear that athletic ability depends on the strengths of bones and muscles.

We do not know exactly how big dinosaur muscles were, because they are not preserved in fossils, so we cannot easily estimate their strengths. We have plenty of dinosaur bones but we cannot usefully measure their strengths, because fossil bones are not made of the same stuff as living ones. Bone consists largely of fibers of the protein collagen reinforced by crystals of the mineral hydroxyapatite. In fossils the protein has decayed and the bone may have been impregnated by other materials seeping through the rock.

Though the composition of bones is changed by the processes of fossilization, their sizes are probably not. We can use the dimensions of the fossil bones to estimate the strengths of the bones in the living dinosaurs, if we assume that dinosaur bone was as strong as bone from modern animals. That assumption seems plausible. Slices of dinosaur bone, examined under a microscope, look very like slices of bone from modern mammals. Samples of bone from the shafts of leg bones of various birds and mammals all have similar properties. (I cannot find any records of strength tests on reptile bone.)

We need to know more about stresses and strains. These words are used in rather vague ways by most people, but are given precise meanings by engineers. Figure 4.1a represents a bar of some material; it does not matter whether it is bone, plastic, rubber or something else. One of its ends is firmly fixed by being embedded in a rigid wall. The grid of squares will help us to see how the bar is deformed when forces act on it.

In figure 4.1b a force is pulling on the free end of the bar, stretching it. The strain is the relative change of length:

$$\text{strain} = \frac{\text{change of length}}{\text{initial length}}$$

Obviously, more force is needed to produce a given strain in a thick bar than in a thin one of the same material. Also more force is needed to break a thick bar than a thin one. This makes it useful to calculate the stress:

$$\text{stress} = \frac{\text{force}}{\text{cross-sectional area}}$$

If the stress reaches a certain limit, called the tensile strength of the material, the bar will break.

In figure 4.1c a force is pushing on the free end of the bar, compressing it. Strain and stress are defined in the same way as before but in this case the strain is negative, because the bar has been made shorter. Also, by convention, the stress is regarded as negative. If the stress reaches a limit, called the compressive strength, the bar will break. The tensile strength of mammal bone is about 160 newtons per square millimeter and the compressive strength is about −270 newtons per square millimeter.

In figures 4.1b and c the forces are in line with the bar but in figure 4.1d a force acts at right angles to the bar, bending it. Notice how the bar is stretched on the outside of the bend and compressed on the inside. The strain and stress are positive on the outside of the bend, and negative on the inside. If the stress on the outside reaches the tensile

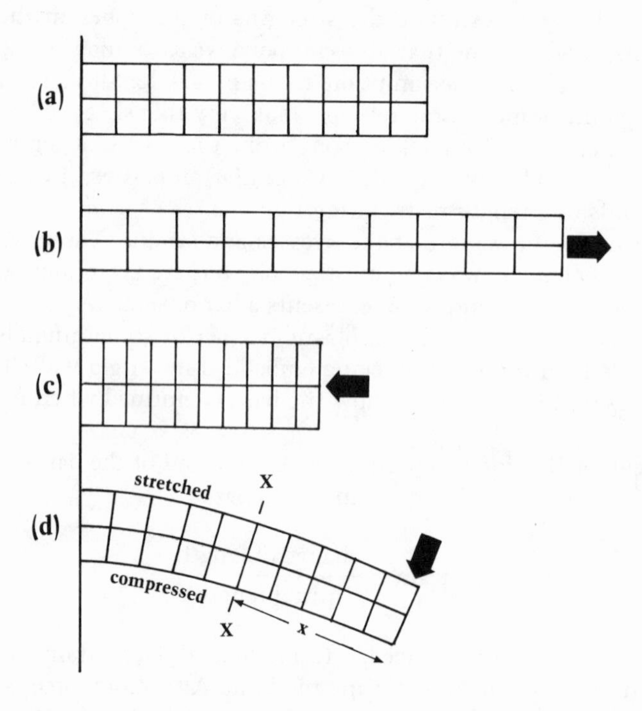

FIGURE 4.1. A bar fixed rigidly at one end, acted on by various forces. These diagrams are explained in the text.

strength, or if the stress on the inside reaches the compressive strength, the bar will break.

In figures 4.1b and c the stresses are the same all along the bar, but in figure 4.1d they are not. The biggest stresses are at the fixed end of the bar, where the force has most leverage. If the force is increased until the bar breaks, it will break where it meets the wall.

Consider one particular cross section of the bar, at XX. This section is at a distance x from the line of the force F. The bending effect of the force at this particular section is described by the bending moment Fx (the force multiplied by the distance). The stresses in this section range from Fx/Z at the top surface of the bar (on the outside of the bend) to $-Fx/Z$ at the bottom surface (on the inside of the bend). In these formulae, Z is a quantity called the section modulus, that depends on the size and shape of the cross section and also on the direction of bending. Its value for a circular cross section is 0.78 times the cube of the radius. Engineering textbooks explain how it can be calculated for other shapes, including irregular ones.

A big section modulus makes a bar hard to break. Figure 4.2 shows some sections that have equal cross-sectional areas, but different section moduli. The tube and the I beam have larger section moduli than the cylindrical rod. Bicycle frames are made of tubes, and girders are given I sections, to make them stronger against bending moments than simple rods would be.

Now we are ready to start thinking about stresses in dinosaur leg bones. Figure 4.3a represents the leg skeleton of a running dinosaur. The foot is pressing down on the ground and the ground is pressing up on the foot. The precise direction of the force may vary, depending on the stage of the stride and on whether the animal is accelerating or decelerating, but we will suppose that the direction is as shown by the arrow. The foot is pivoted at the ankle, so this force tries to rotate the foot counterclockwise about the ankle joint. A balancing force is needed, trying to rotate the foot clockwise. This would be supplied by calf muscles which would probably be arranged as shown in the diagram. (Birds and crocodiles have muscles like this.)

We must think about the balance of forces in more detail. In figure 4.3b two forces are shown acting on the sole of the foot instead of one: a force P parallel to the tibia and a force R at right angles to it. These two forces acting together would have exactly the same effect as the single force shown in figure 4.3a, but it will be convenient to break the force down into these components: it will help us to think about the stresses in the tibia.

In figure 4.3b, the ground exerts forces P and R on the sole of the foot, and the calf muscles exert force M on the foot behind the ankle.

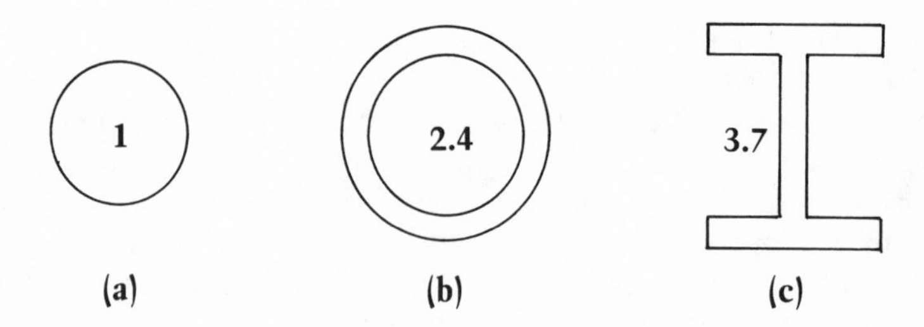

FIGURE 4.2. Possible shapes for the cross section of a beam: (a) is a solid rod, (b) a tube and (c) an I-beam. All have the same cross-sectional area. The numbers are section moduli for bending by vertical forces, relative to the value for the solid rod.

The only other forces on the foot act at the ankle joint, where the tibia presses against the ankle bones. Notice that all the forces shown in figure 4.3b are either parallel to the tibia or at right angles to it. The forces in each of these directions must balance. Forces M and P act upward, parallel to the tibia, so the tibia must press downward with a balancing force $M + P$. Force R presses backward on the sole of the foot so the tibia must press forward at the ankle, with a force R.

Now think about the forces on the lower end of the tibia. The tibia presses downward on the foot with a force $M + P$ so the foot must press upward on the tibia with equal force (figure 4.3c). The tibia presses forward on the foot with a force R so the foot presses backward on the tibia with an equal force. Force $M + P$, acting along the shaft of the tibia, compresses the bone but force R acting at right angles to the shaft of the tibia, bends it.

A cross section of the tibia is picked out by shading in figure 4.3c. This section has area A and section modulus Z, and is at a distance x from the end of the bone. Force $(M + P)$, acting alone, would set up a stress $-(M + P)/A$ in the cross section (remember that compressive stresses are negative). Force R, acting alone, would exert a bending moment Rx and set up stresses ranging from Rx/Z at one edge of the section to $-Rx/Z$ at the opposite edge. The stresses produced by the two forces acting together can be calculated by adding up their separate effects. Thus the total stress ranges from $-(M + P)/A + Rx/Z$ at one edge of the section to $-(M + P)/A - Rx/Z$ at the other.

This analysis tells us that the forces at the ankle would have both a compressing effect on the tibia and a bending effect. It also shows how the stresses could be calculated if the forces on the feet were known. The tibia has been used as an example but similar calculations could be made for the other long bones of the legs.

This could be the starting point for a fearsomely elaborate series of calculations. We could try to reconstruct the precise sequence of movements made by running dinosaurs, perhaps by making animated cartoons. We could calculate the forces that would act on the feet and use these, together with the bone angles taken from the cartoons, to calculate the forces $M + P$ and R on each of the major leg bones. Then, if we knew the dimensions of the bones, we could calculate the stresses in them. We could do this for several different running speeds, calculating the stresses for each. We would find that higher speeds produced bigger stresses and we would be able to estimate the speeds at which the bones would break if the muscles could make the animals run so fast. Maximum running speeds would be less than this, allowing some margin of safety.

Such calculations would raise all sorts of doubts. Did dinosaurs take long strides or short ones? How long did the foot remain on the ground, in each stride? Did they run with their legs relatively straight like elephants or bent like smaller animals? Did they get the foot into a particular position by bending the knee while keeping the ankle relatively straight or by bending the ankle while keeping the knee straight? Were bone stresses at maximum running speeds almost enough to break the bones, or was there a wide margin of safety?

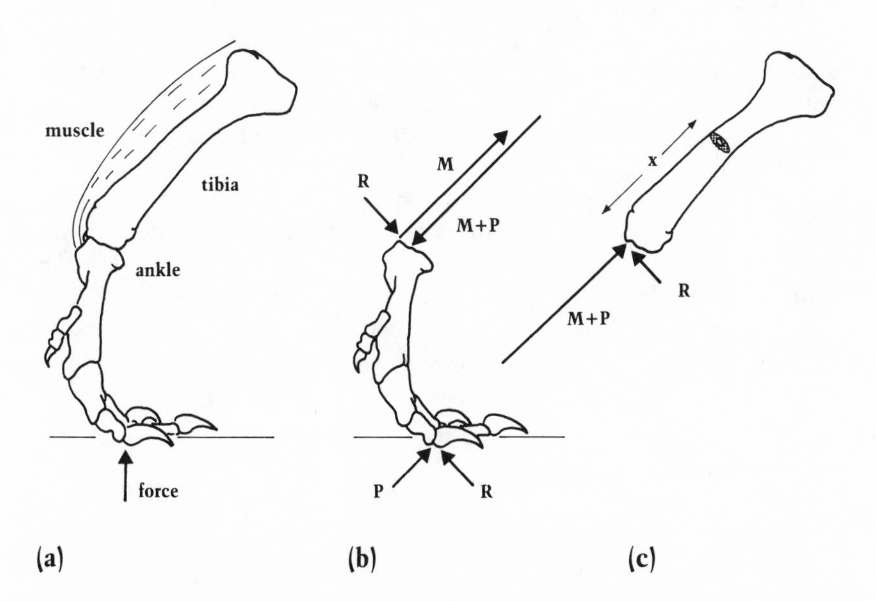

FIGURE 4.3. Forces on the foot and tibia of *Tyrannosaurus*. Diagram (a) shows the leg skeleton, the force on the sole of the foot and the muscles that extend the ankle; (b) shows all the important forces on the foot; and (c) shows forces on the lower end of the tibia.

I faced all these questions in my research on moas, the giant ostrich-like birds that are discussed in chapter 11. Here we avoid facing them directly by making the general assumption, that the movements of dinosaurs and the stresses developed in their skeletons were much the same as for modern animals. We also avoid complicated calculations. I believe that the simple method I am going to describe gives at least as good an indication of the athletic ability of dinosaurs as any other method so far invented.

I must make my assumptions clear. The first assumption is that dinosaurs moved much like modern animals. More precisely, I will assume that the movements of dinosaurs were dynamically similar to those of modern mammals, traveling at the same dimensionless speed. I explained why this assumption seems reasonable, when I discussed dinosaur footprints (chapter 3). Precise dynamic similarity is impossible unless the animals being compared are exactly the same shape, but I will make the best comparisons I can. I will compare sauropods with elephants, which are obviously not the same shape: sauropods have relatively small heads on long necks and elephants have large heads on short necks. Nevertheless, the proportions of the legs are similar (figure 4.4). Both sauropods and elephants have relatively long femurs and short toes, and it seems likely that sauropods moved their legs in much the same way as elephants. However, elephants have big heads and puny tails, so most of their weight is supported by their front legs. Sauropods like *Apatosaurus* had huge hindquarters and tails, so most of their weight must have been supported by their hind legs. I will allow for this difference.

The second assumption is about stresses. I will assume that the thicknesses of dinosaur leg bones were adjusted so that, when the dinosaur ran as fast as it could, the stresses in its leg bones were the same as in the leg bones of modern animals, running as fast as they can. For example, I have calculated that the peak tensile and com-

FIGURE 4.4. Outlines of *Apatosaurus* and an African elephant, with some of the bones drawn in. The outline of the elephant has been traced from a film of fast running. Alexander et al. 1979.

pressive stresses in the humerus of an elephant running fast were about +69 and −85 newtons per square millimeter (about one third of the stresses that would break the bone). I will assume that when sauropods ran as fast as possible the stresses in their bones were about the same. This implies that, if dinosaur bone had the same strength as elephant bone (as is likely), sauropod leg bones had the same margin of safety as elephant leg bones. This assumption will help us to decide whether sauropods were probably more athletic than elephants, or less.

Our calculations would still be complicated, were it not for something very convenient. We can ignore the forces parallel to the bone (M and P in figure 4.3) and consider only the force at right angles to it (R). This may seem odd, because ($M + P$) is usually much larger than R. (Leg muscles attach close to the joints, so their lever arms are short and the forces M that they have to exert are usually much larger than the forces on the feet.) However, forces at right angles to leg bones are much more effective at producing stresses than forces acting along their lengths. This is because the bones are long and relatively thin. It is much easier to break a bone, or a stick, or any other long thin object, by bending it than by compressing it lengthwise.

The most direct evidence we have about the relative importance of compression and bending, in the leg bones of animals, comes from experiments in which strains have been measured in bones of living animals. These experiments used strain gauges, small devices made of metal foil mounted on plastic. Stretching or compressing a strain gauge changes its electrical resistance, so the gauge can be used to measure strain at the surfaces of structures. In surgical operations on horses, sheep, and other animals, strain gauges were glued to the surfaces of leg bones. The skin was sewn up again, but wires from the strain gauges were brought out through the incision so that they could be connected to recording equipment whenever strain was to be measured. Animals recover quickly from the operation and can soon run normally. The operation is so simple and painless that Dr. Hampson, a member of a team doing this kind of experiment, allowed his colleagues to perform the operation on him, attaching a strain gauge to his tibia.

Figure 4.5 shows records from a jumping pony. It had two strain gauges glued opposite each other on a leg bone, one on the front surface of the bone and one on the rear. The records show strains indicating positive (tensile) stresses up to 50 newtons per square millimeter under the front gauge during landing and negative (compressive) stresses up to 60 of the same units under the rear one. The positive stresses almost match the negative ones, indicating that most of the stress was due to bending (as in figure 4.1d) rather than lengthwise compression (as in figure 4.1c). Experiments on various leg bones of horses, sheep and dogs gave sim-

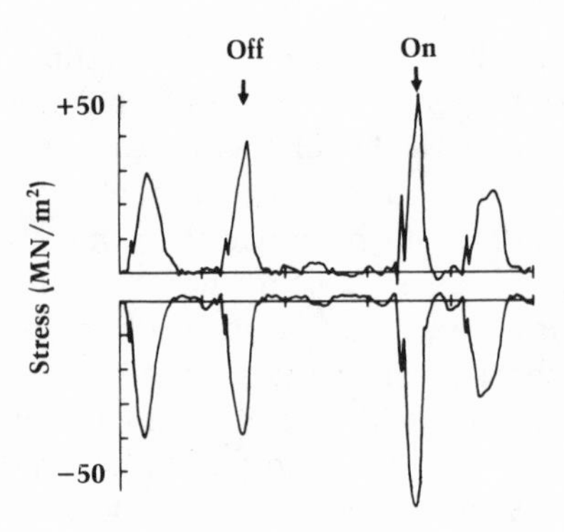

FIGURE 4.5. Stresses calculated from records of strain in the front and rear surfaces of a leg bone (the radius) of a pony, which took off for a jump at "off" and landed at "on." From Biewener, Thomason, and Lanyon 1983.

ilar results in nearly every case. There seems to be a general rule that bending moments are usually much more important than lengthwise compression, in causing stresses in bones.

Think about the example of the elephant humerus. The calculated stresses at the surfaces of the bone were +69 and −85 newtons per square millimeter. They were made up of −8 newtons per square millimeter, in each case, due to lengthwise compression and +77 or −77 newtons per square millimeter due to the bending moment. ($-8 + 77 = 69$ and $-8 - 77 = -85$.) If I had simplified my calculations by considering only the bending moment I would have obtained +77 and −77 instead of −69 and −85, and would not have been far wrong. We will consider only stresses due to bending moments.

The stresses due to bending moments in a cross section of a bone range from $+Rx/Z$ to $-Rx/Z$ (figure 4.3). Suppose that we measure the distance x and the section modulus Z for corresponding sections of the same bone, in two animals. If the animals make dynamically similar movements, their forces R will be in proportion to the weight W that the legs have to support. (If the animals are bipeds, W are the weights of their bodies. If they are quadrupeds, W means the part of the body weight supported by the fore or hind legs, whichever are being considered.) Thus the stresses due to bending moments will be proportional to Wx/Z. The stresses in the two animals will be almost equal, when they make dynamically similar movements, if they have equal values of Wx/Z.

This suggests that we can use Z/Wx (which is Wx/Z turned upside down) as an indicator of athletic ability. The bigger Z/Wx is, the less stress will act, in any particular activity. Large values of Z/Wx mean that strenuous activities are possible without setting up dangerous stresses in the bones. We can estimate the athletic prowess of dinosaurs by calculating values of Z/Wx and comparing them with values for modern animals, whose athletic ability we know. Whichever animal has the larger values of Z/Wx is likely to be the more athletic.

We are going to apply this formula to quadrupedal dinosaurs, so we need to know how to measure W, the weight supported by the fore or hind legs. The measurement is easily made for living animals, if a force plate is available. When the animal stands with all four feet on the plate, the vertical force registered by the plate is the total weight of the animal. When it stands with its fore feet on the plate and its hind feet on the floor alongside, the force registered is the weight supported by the fore feet. If the animal is too heavy for an ordinary force plate, a weighbridge of the kind used for weighing trucks can be used instead. Measurements on a weighbridge showed that an elephant carried 58 percent of its weight on its front legs and 42 percent on its hind legs.

We cannot get living dinosaurs to put on weighbridges so we have to use models, and there is a problem about doing that. The models I used were solid plastic, of the same density throughout, but different parts of the living dinosaur had different densities. Bone is about twice as dense as flesh and guts (i.e. the same volume is about twice as heavy). If a dinosaur had a lot of bone at its front end and very little in its tail (for example) its fore feet would have to support a bigger fraction of the total weight than they do in a plastic model. The bones of dinosaurs were not concentrated near one end, but distributed all through the body, so we probably do not need to worry about the model not having dense bones. Also if the front half of the dinosaur were filled with air, the fraction of body weight carried by the fore feet would be less than for a solid model. There must have been quite a lot of air in the lungs, which would have been in the front half of the body, and it seemed necessary to take account of it. I therefore bored holes through the models where the lungs would have been. Experiments on caimans and on large mammals such as elands and camels had shown that their lungs occupy about 8 percent of the volume of the body. I assumed that the same would be true of dinosaurs and adjusted the sizes of the holes so as to reduce the mass of each model by 8 percent.

Figure 4.6 shows how I did the experiment, using a letter balance instead of a weighbridge. When I arranged the dinosaur model as shown, the balance read 51 grams. I removed the model leaving the rubber pad on the pan, and the reading fell to 5 grams. The mass of the model

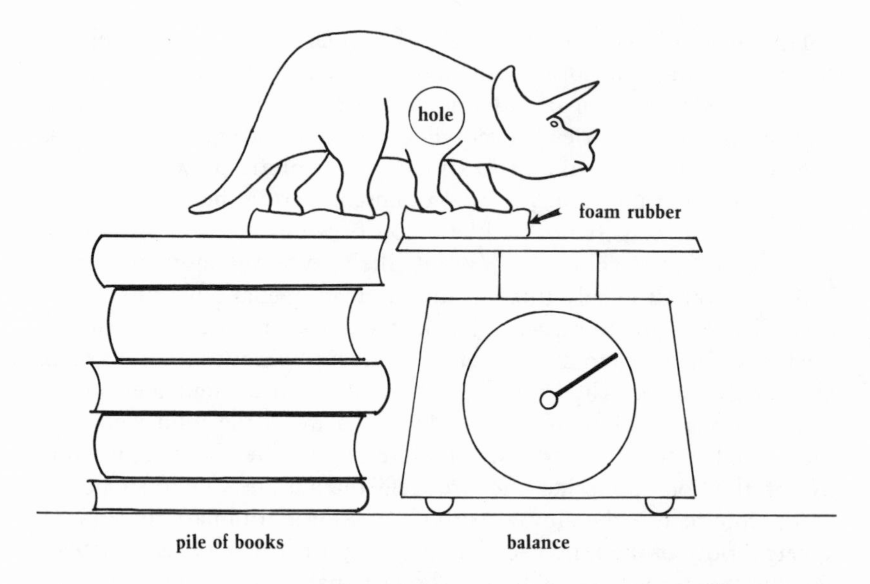

FIGURE 4.6. An experiment to discover how much of the weight of a dinosaur was supported by its fore legs. The circle on the model's chest is a hole bored to represent the lungs.

was 85 grams. Therefore, the fraction of the model's weight supported by the fore legs was $(51 - 5)/85 = 0.54$.

The pads of foam rubber had an important function. The model was rigid, so only three of its feet would probably rest on the ground when it stood on a rigid surface. Similarly a table on a flat floor rests on just three of its four feet unless the lengths of its legs are very precisely matched. The pieces of foam rubber ensured that the weight was shared between all four feet.

The results of the experiments are shown in table 4.1. It seems that *Brachiosaurus* supported almost half its weight on its fore feet, but that the other sauropods, *Diplodocus* and *Apatosaurus*, supported most of their weight on their hind feet. (The *Apatosaurus* values come from a homemade model and are probably less accurate than the rest.) *Stegosaurus* also carried most of its weight on its hind feet but *Triceratops* carried more weight on its fore feet.

The living animals in the table are not the ones we would most like to know about. The elephant is an Indian one, but my measurements of elephant bones are from an African elephant. I will assume that the percentages of body weight carried by the fore and hind feet are the same for both species. I will also assume that the percentages for buffalo are about the same as for a horse.

At last we are ready to calculate Z/Wx, the indicator of athletic ability. My values for the section modulus Z and the distance x come from measurements of bones, both from modern animals and from dinosaur fossils. The sections that I measured were usually about half way along the bones. I knew the masses of the modern animals because the bones came from weighed carcasses, and I had calculated the masses of the dinosaurs from the volumes of models (table 2.2). I took the fractions of body weight carried by the fore and hind feet from table 4.1, and so was able to calculate the weight W carried by each pair of feet.

Notice that I use dinosaur masses calculated from the volumes of models, that is masses taken from the "Colbert" and "Alexander" columns in table 2.2. It would have been wrong to use the "Anderson" masses for this job because they were calculated from bone dimensions. They assume a particular relationship between body mass and the dimensions (and so the strengths) of the bones. If we used them to assess the athletic ability of dinosaurs, by comparing the strengths of bones to the loads they have to carry, our argument would be circular.

The resulting values of Z/Wx are shown in table 4.2. Notice that the values for elephant bones are all much lower than the values for corresponding bones of buffalo. The comparison is not a particularly good one, because the relative lengths of the leg bones of elephants and buffaloes are very different, so their movements can never be precisely dynamically similar. However, the values do indicate correctly that elephants are less athletic than buffaloes. Most mammals walk at low

TABLE 4.1. Percentages of body weight supported by the fore and hind feet of various quadrupeds.

	PERCENTAGES OF BODY WEIGHT SUPPORTED BY	
	Fore Feet	*Hind Feet*
Indian elephant	58	42
Horse	59	41
Diplodocus	22	78
Apatosaurus	30	70
Brachiosaurus	48	52
Triceratops	54	46
Stegosaurus	18	82

speeds, trot at intermediate speeds, and gallop at high speeds. Elephants generally do not trot but use an alternative running gait, the amble, with a different sequence of foot movements (figure 4.4). Their fastest gait is a slow amble, and they seem unable to gallop. Buffaloes can gallop, though not very fast. Elephants do not jump and are often kept in their enclosures in zoos by a narrow ditch instead of a fence. I have no data on buffalo jumping but I have film of an impressive jump by an eland (an antelope of about the same mass).

The quadrupedal dinosaurs in table 4.2 had legs with elephant-like proportions, so comparisons with elephants should be revealing. The value of Z/Wx for *Diplodocus* femur is even lower than for the elephant, which suggests that *Diplodocus* was even less athletic than elephants: it could walk but probably could not run. The values for the humerus, femur, and tibia of *Apatosaurus* are all close to the values for elephant, which suggests that *Apatosaurus* was about as athletic as an elephant, able to run but not to gallop or to jump. The values for *Triceratops* are close to those for buffalo, suggesting that *Triceratops* may have been able to gallop.

The largest modern galloping animal seems to be the White rhinoceros, which is said to reach 3 tonnes, about half as much as a really large African elephant. I have drawn the rhinos in figure 4.7 from a film of a rhinoceros galloping. It was being chased by a vehicle across

TABLE 4.2. Values of Z/Wx (the quantity used as an indicator of athletic ability) for leg bones of dinosaurs and modern animals.

	Body Mass (Tonnes)	*Values of Z/Wx (Square Meters per Giganewton)**		
		Femur	*Tibia*	*Humerus*
African elephant	2.5	7	9	11
Buffalo	0.5	22	27	21
Diplodocus	12–19	3–5		
Apatosaurus	34	9	6	14
Triceratops	6–9	15–21		12–20
Ostrich	0.04	44	18	
Man	0.06	15	15	
Tyrannosaurus	8	9		

*A giganewton is one billion (10^9) newtons.

FIGURE 4.7. Outlines traced from a film of a White rhinoceros galloping, with drawings of *Triceratops* galloping based on the same film.

a very large zoo enclosure, and reached a maximum speed of 7.5 meters per second (17 miles per hour). If *Triceratops* could gallop, it presumably galloped like a rhino.

There are no really large modern bipeds to compare with the large bipedal dinosaurs. The largest are ostriches and humans, and I have put data for both in table 4.2. The man whose bones I measured had been ill and inactive for a few years before he died so his bones may have become thinner and weaker than when he was healthy. Even so, Z/Wx for his femur was higher than for *Tyrannosaurus*, but the value for ostrich femur was *much* higher. These comparisons suggest (unsurprizingly) that *Tyrannosaurus* was less athletic than ostriches and people. The *Tyrannosaurus* value is close to the elephant one, which again suggests that *Tyrannosaurus* could not have run very fast. The proportions of *Tyrannosaurus'* hind legs are unlike those of ostriches, people or elephants, so none of these comparisons are really satisfactory, but they seem to be the best possible.

Whether they were strong enough for fast running or not, dinosaur skeletons must have been strong enough to support the animals when they copulated. The male rhino in figure 4.8a seems to be imposing a terrific load on the female's back. Male elephants also mount females to copulate like this, but dinosaurs must have done it rather differently because their big tails would get in the way. Reptiles do not have the

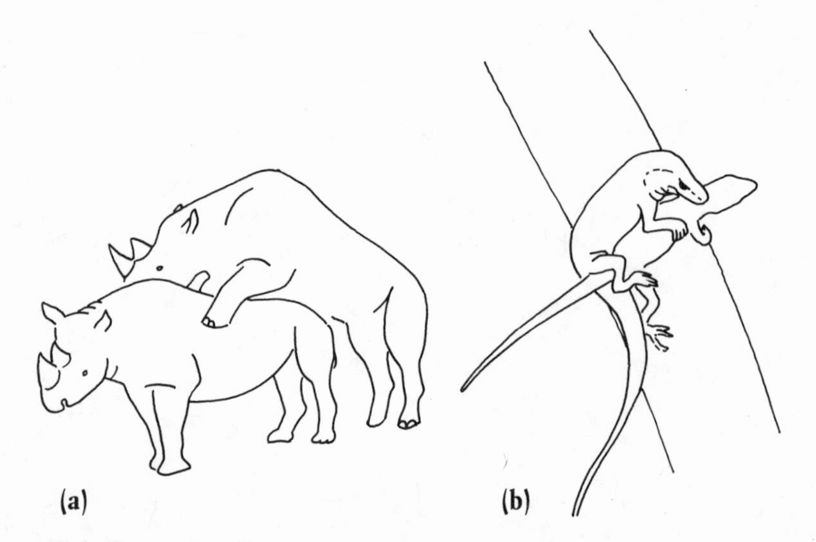

FIGURE 4.8. (a) Black rhinoceros copulating, from a photograph; (b) *Anolis* lizards copulating on a branch of a tree, from Greenberg and Noble 1944.

sexual orifice separate from the anus, like humans and other mammals, but have the two combined to form a single opening called the cloaca. This is on the underside of the tail, a short distance behind the legs. To copulate, the male must press his cloaca against the female's. That may sound impossible to do from behind, but lizards manage by twisting their tails together (figure 4.8b). Dr. Beverley Halstead, a British paleontologist, has suggested that dinosaurs also may have copulated like this, with the male lifting one hind leg and putting it over the female's back.

If they did, bigger loads would act on their hind legs than in normal standing. When the male lifted one hind leg, the other would have to carry twice the standing load. If he rested his leg on the female's back, her hind legs would have to carry an increased load. However, the loads would be no bigger than in walking, when peak forces on the feet were probably about twice as much as in standing. (I explained at the beginning of this chapter why peak forces on human feet are twice as much in walking as in standing.) If dinosaurs were strong enough to walk they were also strong enough to copulate. They were presumably strong enough to do both.

This chapter has told a complicated story. Here, briefly, is how it went. The faster an animal runs or the more athletically it behaves the bigger, in general, are the forces on its legs. It is harder for large animals than for small ones to withstand these forces because doubling the length of an animal without changing its proportions makes it eight times heavier but only four times stronger. This is why elephants are less athletic than gazelles.

I used the dimensions of dinosaur bones to assess athletic ability. I assumed that the running movements of dinosaurs were much like those of modern animals and that their skeletons were proportioned so that the stresses in them, in the most strenuous activities, were the same as for modern animals. I showed that, for rough calculations of stress, only forces that set up bending moments in the bones need be considered. This led to the definition of the quantity Z/Wx, that takes account of the dimensions of a leg bone and of the weight supported by the legs. The larger its values of Z/Wx, the more athletic an animal is likely to have been.

I compared quadrupedal dinosaurs with elephants. The values of Z/Wx suggested that *Apatosaurus* was about as athletic as an elephant, able to run slowly but not to gallop. *Diplodocus* was less athletic but *Triceratops* seems to have been more athletic, and may have galloped like a rhinoceros. *Tyrannosaurus* was much less athletic than ostriches or people, the largest modern bipeds.

Finally (and rather obviously), dinosaurs were strong enough to copulate.

Principal Sources

I have based this chapter on Alexander (1985) but have used a new method to estimate the weights supported by the fore and hind feet of dinosaurs. Consequently, some of the numbers given in the chapter are slightly different from those in the original paper. The new method is not necessarily better than the old one but it is more direct, and so easier to explain. The information about elephants and buffaloes comes from Alexander et al. (1979). The data from strain gauges glued to bones of living animals are from Biewener, Thomason, and Lanyon (1983).

Alexander, R. McN. 1985. Mechanics of posture and gait of some large dinosaurs. *Zoological Journal of the Linnean Society* 83:1–25.

Alexander, R. McN., G. M. O. Maloiy, B. Hunter, A. S. Jayes, and J. Nturibi, (1979). Mechanical stresses in fast locomotion of buffalo (*Syncerus caffer*) and elephant (*Loxodonta africana*). *Journal of Zoology* 189:135–144.

Biewener, A. A., J. Thomason, and L. E. Lanyon, 1983. Mechanics of locomotion and jumping in the forelimb of the horse (*Equus*): *In vivo* stress developed in the radius and metacarpus. *Journal of Zoology* 201:67–82.

Greenberg, B. and G. K. Noble, 1944. Social behavior of the American chameleon (*Anolis carolinensis* Voigt). *Physiological Zoology* 17:392–439.

V

Dinosaur Necks and Tails

ALMOST ALL dinosaurs had long tails and some also had long necks. In an extreme case, *Diplodocus* had a neck 7 meters (23 ft) long (including the small head), a tail about 14 meters (46 ft) long and body only 5 meters (16 ft) long. This chapter asks how dinosaurs used long necks and tails.

I will start with necks. The longest are those of sauropods, which include the largest of all dinosaurs. Before the calculations about bone strength had been done, many people doubted whether the biggest sauropods could have supported their weight on land. They supposed that they must have lived in lakes, buoyed up by the water (figure 5.1). Their long necks might have served as snorkels, enabling them to breathe while standing on the bottom in deep water.

The sauropods in figure 5.1 look like *Diplodocus.* If they are, their lungs are 6 meters (20 ft) or more below the surface of the water. Their necks are long enough to reach the surface but it would be very hard for them to breathe air in, because the lungs would have to be expanded against a pressure difference of 6 meters of water. The snorkels used by human divers are only about 30 centimeters (1 ft) long. The dinosaurs in figure 5.1 would have needed enormous chest muscles, to breathe in.

It now seems clear that even the largest sauropods had legs strong enough for walking on land, so there is no need to imagine them living submerged in lakes. Indeed, it has been argued that they probably lived on dry land, keeping clear of marshy places where they could easily have got bogged down because of the enormous pressure on their feet (chapter 3). If they lived on land, as is now generally believed, the long neck cannot have been a snorkel. We must look for some other function.

The shape of *Brachiosaurus* suggests that it lived like a giraffe, eat-

FIGURE 5.1. Snorkeling sauropods, from Gregory 1951.

ing leaves from tall trees. Not only was its neck very long, but the fore legs were longer than the hind (a difference from other sauropods), so its head could apparently be raised to the remarkable height of 13 meters (43 ft). It was much taller than any giraffe (figure 1.1).

If *Brachiosaurus* carried its head so high, its brain would be about 8 meters (26 ft) above its heart. The heart would have to pump blood at very high pressure, to get it to the brain. If blood failed to get to the brain even for a short time, the animal would collapse unconscious. A fainting *Brachiosaurus* would come down with a tremendous crash.

Blood pressure is conventionally measured in millimetres of mercury because doctors used to use mercury manometers to measure human blood pressure. It will be more convenient here to use meters of water: a pressure of one meter of water equals 74 millimeters of mercury. Modern reptiles pump blood out of their hearts at pressures that are 0.5 to 1.0 meters of water above the pressure of the surrounding tissues. This pressure difference is needed mainly to drive blood through the capillaries, the fine blood vessels that permeate every tissue of the body. *Brachiosaurus* would have needed an additional pressure of 8 meters of water to raise the blood to the brain, making a total of 8.5 or more meters of water. (Strictly speaking, the additional pressure would be 8 meters of *blood,* not water, but this makes little difference because the densities of blood and water are only slightly different.)

The calculated total blood pressure, 8.5 meters of water, is much larger than the blood pressure of any modern animal. Most mammals, including ourselves, pump blood from their hearts with pressures of 1.5 to 2.0 meters of water, and even giraffes do it with pressures no higher than 4.3 meters of water. Giraffes need much less blood pressure than *Brachiosaurus* because their necks are so much shorter, carrying the brain only 3 meters above the heart.

You might think that the blood systems of giraffe and dinosaur necks could work like siphons, which do not need high pressures to drive liquids through them. However, you need a rigid tube to make a siphon because when it is working, the pressure inside it, near the top, is less than the pressure outside. Veins have flexible walls that would collapse if the pressure inside fell, so they cannot work as siphons.

It would be a mistake to argue that a blood pressure of 8.5 meters of water would be impossible, because it is so much higher than the blood pressures of modern animals. If giraffes were unknown, that kind of argument would lead to the conclusion that giraffes were impossible. However, the blood pressure calculated for *Brachiosaurus* does seem remarkable, and a very muscular heart would have been needed to produce it.

Brachiosaurus had long fore legs but *Diplodocus* and *Apatosaurus* had relatively short ones, and look much less like giraffes. In 1978 Dr. Robert Bakker suggested that they also may have fed from tall trees, rearing up on their hind legs to gain extra height (figure 5.2e,f). It may seem ludicrous to suggest that these enormous animals could manage such gymnastics, but the idea deserves careful consideration. Remem-

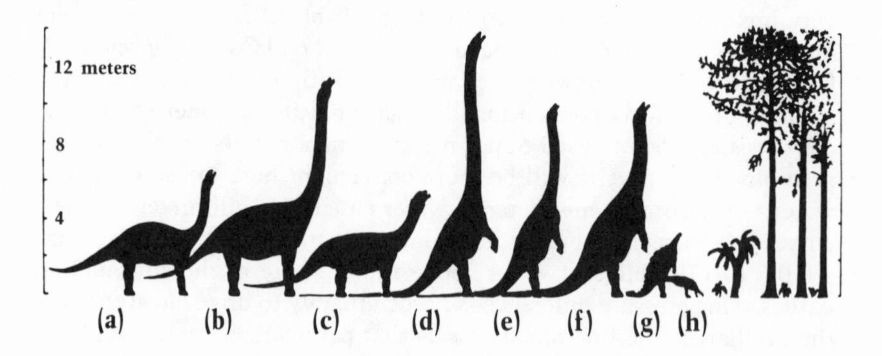

FIGURE 5.2. Possible postures of some dinosaurs when browsing on foliage: (a) *Haplocanthosaurus;* (b) *Brachiosaurus* (not a very large one); (c) *Camarasaurus;* (d)*Barosaurus;* (e) *Diplodocus;* (f) *Apatosaurus;* (g) *Stegosaurus;* and (h) (feeding from the ground) *Camptosaurus.* From Bakker 1978. Reprinted by permission. Copyright © 1978 Macmillan Magazines Ltd.

ber that circus elephants can be trained to balance on their hind legs, and that the calculations in chapter 4 suggest that *Apatosaurus* may have been as athletic as elephants. The stresses in the bones, while standing on the hind legs, would probably be less for *Apatosaurus* than for elephants. This is because in its normal quadrupedal position *Apatosaurus,* unlike elephants, carried most of its weight on its hind legs (table 4.1). Standing on the hind legs alone would not increase the load on them very much.

It might be easy for *Apatosaurus* to support itself, once it was standing on its hind legs, but could it get itself into that position? It would have to get its center of gravity over its hind feet, to take the load off its fore feet. To discover whether that would be difficult we need to know where the center of gravity was. We already know that *Apatosaurus* and *Diplodocus* carried most of their weight on the hind legs, which implies that the center of gravity was nearer the hind legs than the fore. We could calculate its position from the data in table 4.1 but we can discover it more directly, by a simple experiment.

For this I used the same models as in chapter 4, the solid plastic ones with holes bored to represent the air-filled lungs. Their centers of gravity should be in approximately the same positions as in the living dinosaurs. I suspended each model by a thread tied to its head and photographed it in side view (figure 5.3a). The model was hanging motionless, in equilibrium, so its center of gravity must have been in line with the thread, somewhere on the line AB. Then I suspended the model from its back and took another photograph (figure 5.3b). The center of gravity must again have been in line with the thread, on the line CD. Finally, I superimposed the two photographs, making the outlines of the model coincide (figure 5.3c). The center of gravity was both on AB and on CD: it must have been at the intersection of the two lines.

Figure 5.4 shows centers of gravity located in this way. *Diplodocus* standing in the position shown had its center of gravity over the left hind foot. If the animal had moved its right hind foot forward and set it down beside the left one, both hind feet would have been under the center of gravity and it would have been easy for it to rear up on its hind legs.

Diplodocus could apparently have reared up easily because its long, heavy tail counterbalanced the front part of the body and brought the center of gravity well back. *Brachiosaurus* had a shorter tail and heavier fore quarters, so its center of gravity was further forward. It would have been harder for it to rear up and, as far as I know, neither Bakker nor anyone else has suggested that it did. With its long front legs and neck it could feed from great heights without rearing up.

Bakker suggested that *Stegosaurus* reared up on its hind legs to feed, and it seems likely that it could have done: its center of gravity was well back, near the hind legs. *Triceratops* had its center of gravity much further forward (figure 5.4) and might have found it difficult to rear up. However, elephants have their centers of gravity well forward, and can rear up. (Table 4.1 shows that the centers of gravity of elephants are far enough forward to put most of the weight on the front feet in normal standing.)

It is generally assumed that long-tailed sauropods such as *Diplodocus* and *Apatosaurus* walked around with their necks nearly horizontal. Most drawings show them in that position, as do the mounted skeletons in museums. We will think about the problem of supporting a long, heavy, horizontal neck.

I am going to suggest that these necks were supported in the same way as the necks of horses, cattle, and their relatives. These animals have a thick ligament called the ligamentum nuchae running along the backs of their necks (figure 5.5). Unlike most other ligaments it consists mainly of the protein elastin, which has properties very like rubber. It can be stretched to double its initial length without breaking and snaps back to its initial length when released.

The ligamentum nuchae is stretched when the animal lowers its head

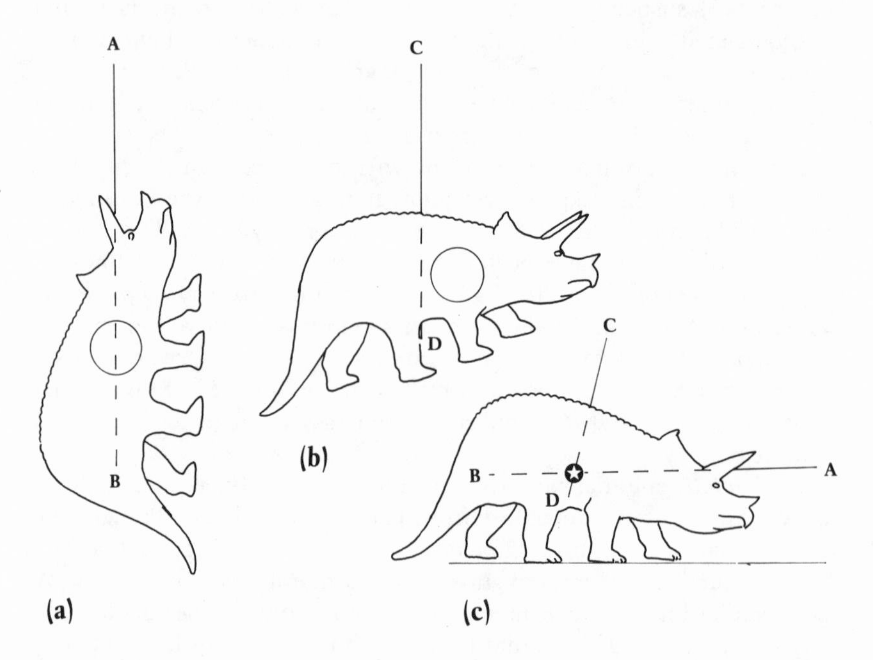

FIGURE 5.3. Diagrams showing how to locate the center of gravity of a model. They are explained in the text.

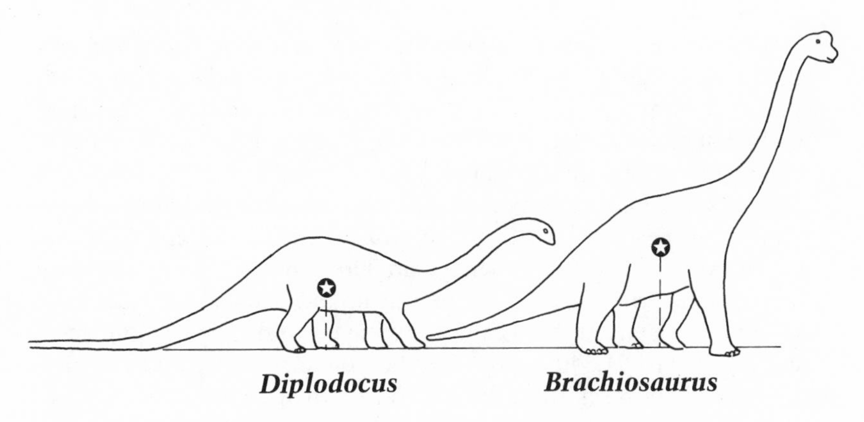

FIGURE 5.4. Outlines of dinosaurs, showing the positions of their centers of gravity.

to drink or graze, and shortens again when the head is raised. In experiments with deer carcasses, my colleagues and I found that the ligament was 1.4 times its slack length when the head was raised to the position of figure 5.5a, and almost twice its slack length when it was lowered to the position of figure 5.5b. Notice that the ligament was stretched even when the head was high: I doubt whether a deer can get into a position that allows the ligament to shorten to the point of going slack. If you cut the ligament in a dissection the cut ends spring apart, as if you had cut a stretched rubber band.

Thus the ligamentum nuchae is taut in all normal positions of the neck. Its tension helps to support the weight of the head and neck. The tension and the supporting effect are greatest when the head is lowered (as in figure 5.5b) but even in this position the tension is not by itself enough to give all the necessary support: some tension in neck muscles is needed as well.

The ligamentum nuchae is a feature of hoofed mammals but something rather similar is found in birds, which are much more closely related to dinosaurs. Instead of a continuous ligament running the whole length of the neck they have a series of short ligaments connecting each neck vertebra to the next (figure 5.6a). These ligaments consist largely of elastin, like the ligamentum nuchae. They are stretched when the head is lowered, and shorten again when it is raised.

Figure 5.6c shows a bird vertebra in front view. The centrum is the main body of the vertebra, connected to the vertebrae in front and behind by intervertebral discs. Above it is a hole for the nerve cord and above that again is the neural spine. The elastic ligaments connect the front of one neural spine to the back of the next.

Diplodocus and *Apatosaurus* have neck vertebrae with V-shaped neural spines (figure 5.6d). I suggest that the V was filled by an elastin ligament that ran the whole length of the neck and back into the trunk. This ligament would have helped to support the neck while allowing the dinosaur to raise and lower its head.

I made some experiments to find out whether the idea was feasible. I cut off the head and neck of the *Diplodocus* model and measured their volume (as I had done with complete models in chapter 2), and calculated that the mass of the real head and neck would have been 1,340 kilograms. Their weight was this mass multiplied by the acceleration of gravity, 1,340 × 10 = 13,400 newtons. I suspended the amputated head and neck from threads to find their center of gravity, and have shown the weight acting at the center of gravity in figure 5.6b.

We can only guess how thick the ligament was, but it seems likely that it projected above the tops of the neural spines, as shown in figure 5.6d. If so its center line, at the base of the neck, was about 0.42 meter above the center of the centrum.

In figure 5.6b the weight is pulling the neck counterclockwise about the joint at its base and the tension in the ligament is pulling it clockwise. The weight acts 2.2 meters from the joint and the ligament tension 0.42 meters from it. By the principle of levers, the tension that would be needed to balance the weight is 2.2 × 13,400/0.42 = 70,000

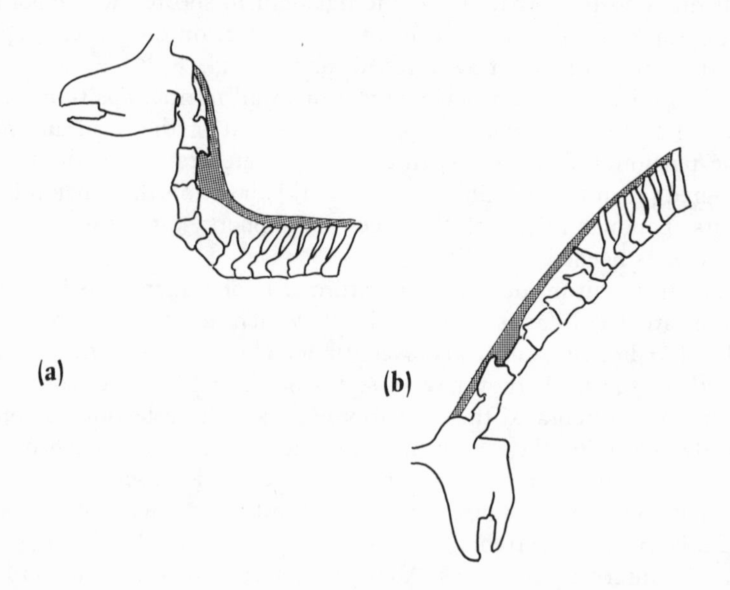

FIGURE 5.5. The neck of a Roe deer in the alert position and lowered for feeding. The skeleton and the ligamentum nuchae (stippled) are shown.

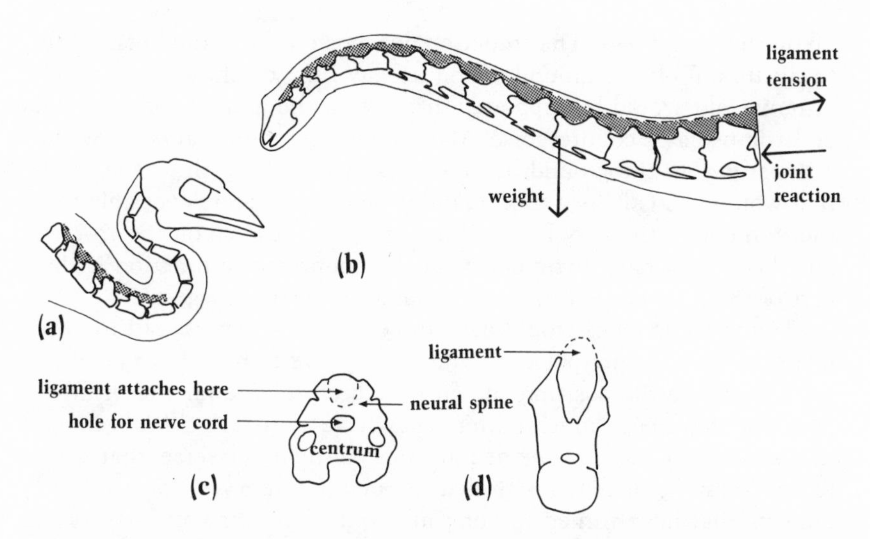

FIGURE 5.6. Necks of (a) a turkey and (b) *Diplodocus,* showing the vertebrae and the elastic ligaments. The forces that acted on the neck of *Diplodocus* are also shown. (c) and (d) are front views of neck vertebrae of an ostrich and of *Diplodocus.*

newtons (7 tonnes force). The third force shown in the diagram is the force in the joint itself, where one centrum presses on the next.

The calculated tension may seen enormous, but the ligament was very thick. If it was as thick as in the diagram, its cross-sectional area was 40,000 square millimeters and the stress in it, for a force of 70,000 newtons, would be 1.8 newtons per square millimeter. This is more than the stress in the ligamentum nuchae of a deer with its head down (about 0.6 newtons per square millimeter), and would be enough, or nearly enough, to break ligamentum nuchae. However, this stress would act only if the ligament supported the neck without any help from muscles. If neck muscles took some of the load (as they do in birds) the stress in the ligament would be less. The suggestion of an elastin ligament seems feasible.

Remember that the ligament is only a guess, based on the structure of sauropod vertebrae and comparisons with modern animals. We do not know whether it existed, but the calculations seem to show that if it did it could have done a useful job. Only some sauropods had V-shaped neural spines that could have housed a continuous ligament but other dinosaurs may have had separate ligaments connecting each neck vertebra to the next, as in birds.

Now we will think about tails. We have already seen how a long tail may balance the front part of the body, enabling some dinosaurs to rear up on their hind legs. That function depends on the animal being able to lift its tail off the ground. If the tail lay limp on the ground, much of its weight would be supported directly on the ground and it would be little use as a counterpoise. Many drawings of dinosaurs show the tail trailing on the ground but few sets of footprints show the mark that would be made by a trailing tail (chapter 3). It seems possible that the thin end of the very long tail of *Diplodocus* trailed on the ground, but this would have little effect on the counterpoise function if this part of the tail was as thin as the slender vertebrae suggest.

Though *Diplodocus* would have to be able to stiffen its tail to raise it for use as a counterpoise, it would also have to be able to bend its tail to get into the position shown in figure 5.2. The tails of ornithopods may have been much stiffer. They have a crisscross arrangement of rods, on either side of the neural spines of their vertebrae, that seem to be ligaments or tendons turned to bone (figure 5.7). (You will find tendons that have turned to bony material whenever you eat turkey, as hard strips embedded in the meat of the lower leg.) Whenever complete skeletons of ornithopods are found, the tail is fairly straight (figure 5.7), suggesting that it was indeed stiff.

How stiff the tail was would depend on whether the rods were ligaments or tendons. Ligaments connect bone to bone, and if the tail vertebrae were connected by rigid bony ligaments the tail would be very stiff indeed. Tendons connect muscles to bones, and if the rods were tendons the tail could have bent up and down a little as its muscles shortened and lengthened. The rods look like tendons to me. Modern mammals have similarly arranged (but non-bony) tendons in their tail muscles.

Bony ligaments or tendons are most prominent in the duck-billed

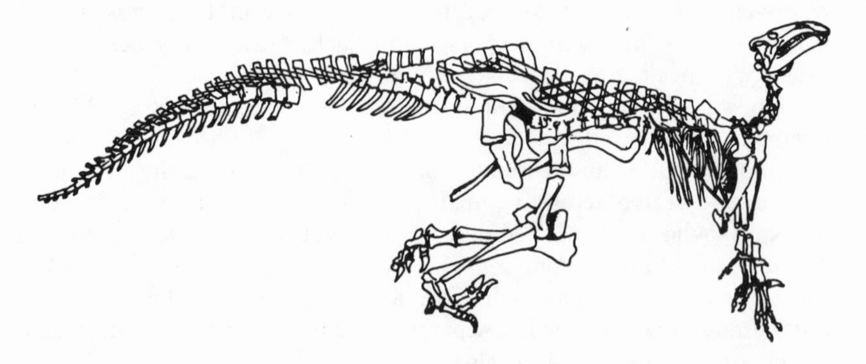

FIGURE 5.7. Skeleton of *Iguanodon*, in the position in which it was found. Notice the crisscross tendons in the back and tail. From Norman 1980.

FIGURE 5.8. An ostrich, a man, and a small bipedal dinosaur (*Hypsilophodon*) running, showing the positions of their hips and their centers of gravity: ○ hip; ● center of gravity.

dinosaurs but have also been found over the hips of *Triceratops,* in the tail of the theropod *Deinonychus* and in several other dinosaurs.

The tail must have had a major effect on the running movements of bipedal dinosaurs. Figure 5.8 compares a dinosaur with two modern bipeds, a bird and a human. The dinosaur, with its long heavy tail, has the center of gravity close to the hips, but the bird, with only a tuft of feathers for a tail, has its center of gravity well in front of the hips. The human has no tail but the center of gravity is close to the hips because the trunk is erect. If these bipeds are not to fall over, the average positions of their feet, while on the ground, must in each case be under the center of gravity. Each animal sets the foot down in front of the center of gravity and does not lift it until it is behind the center of gravity. The bird manages by holding its thigh almost horizontal and swinging the leg from the knee. It seems likely that bipedal dinosaurs moved their legs more like people than like birds, because of the position of the center of gravity.

The tails of kangaroos are thick and heavy, but they are flexible, and they swing up and down as the animal hops. If flexible tails are suitable for kangaroos, why should stiff ones have evolved in bipedal dinosaurs? The answer may depend on the difference between running and hopping and on a basic principle of mechanics.

I will use an example from gymnastics to explain the principle of conservation of angular momentum. A gymnast on a trampoline can set his body spinning, while in mid air, by moving his arms. By swinging his arms to the left he makes his body spin to the right. A rotating object has a property called angular momentum that depends on the masses of its parts and on its rate of rotation. The principle says that the gymnast cannot change his total angular momentum without pushing on something, but he can set his arms rotating in one direction and his trunk in the other so that the angular momentums in the two directions cancel out.

As a kangaroo flies through the air in a hop, it swings its legs forward ready for the next landing. In figure 5.9 its legs have to be given counterclockwise angular momentum so its trunk and tail must get matching clockwise momentum. If the tail were rigid the body would rock up and down quite a lot (through 13–18°, according to my calculations). Actually the tail bends so that it swings up and down through a large angle and the trunk through a much smaller one. These tail movements probably cost the animal very little energy, because they do not have to be powered by muscles. The tail vibrates passively because its long (non-bony) tendons make it springy.

If the trunk rocked a lot, with the head rocking with it, it might be difficult for the animal to keep watch for danger as it hopped. The springy tail may give the kangaroo an advantage, by reducing the rocking movements of the trunk. A dinosaur that ran rather than hopped would not need such a mechanism because, in running, one leg swings back while the other swings forward and there is little tendency for leg movements to rock the body. A springy tail may be best for a hopper but a stiff one seems fine for a runner.

The tails of some dinosaurs may have served as weapons, as well as for balance. It has often been suggested that the long tapering tails of *Diplodocus* and *Apatosaurus* would have made formidable whips. They could have been used to strike a predator, and I wonder whether they could also have been used to make a terrifying noise. When a circus ringmaster cracks his whip, he flicks it so as to make its tip move supersonically. Is it too wild a speculation to wonder whether *Diplodocus* could crack its tail?

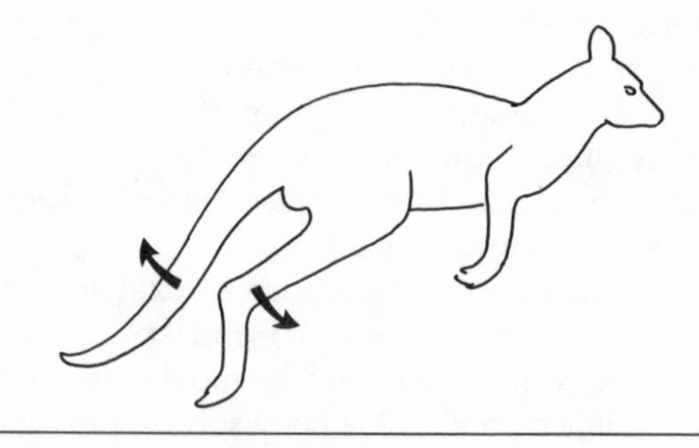

FIGURE 5.9. Diagram of a kangaroo hopping.

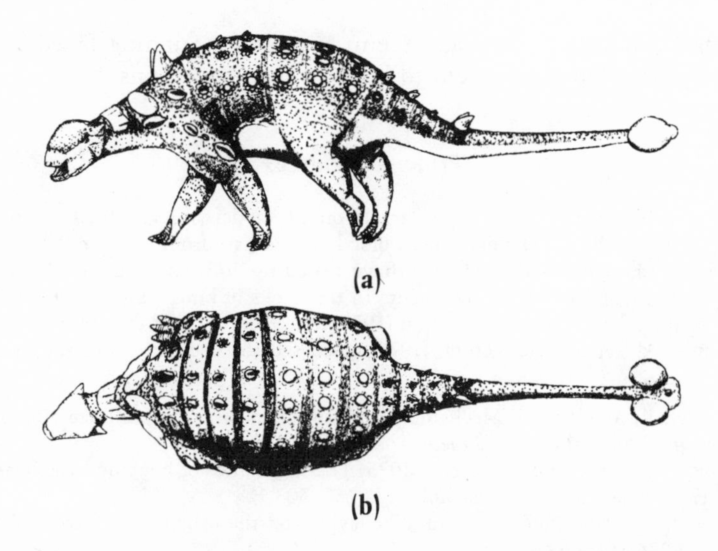

FIGURE 5.10. *Euoplocephalus* or *Dyoplosaurus*, an ankylosaur with a club on its tail (a) in side view and (b) in top view. Length six meters. From Carpenter (1982).

There are other dinosaur tails that seem more obviously to have been weapons. *Stegosaurus* had sharp spikes on its tail up to half a meter (20 in) long. Some of the ankylosaurs had big lumps of bone at the ends of their tails and presumably used them as clubs (figure 5.10).

This chapter has told a series of short stories. Sauropods are unlikely to have used their long necks as snorkels. Very large breathing muscles would be needed for snorkeling at any substantial depth, and in any case it seems unlikely that they lived in water. It seems more likely that they lived on land, raising their necks to feed from high branches. Their hearts would then have had to pump blood out at high pressure, to get it to the brain. *Brachiosaurus* probably fed like a giraffe but *Diplodocus* and similar sauropods may have reared up on their hind legs to get their heads high. Their long, heavy tails would probably have made it easy for these sauropods to get their hind feet under their center of gravity. The long necks of *Diplodocus* and *Apatosaurus* may have been supported by an elastin ligament running through the V-shaped notches in their neural spines.

The tails of duck-billed dinosaurs had bony tendons or ligaments alongside the neural spines and seem to have been stiff. The effect of the tail on the position of the center of gravity probably made bipedal dinosaurs move their legs more like people than like birds. The springy tails of kangaroos may help to prevent the body from rocking too much during hopping but stiff tails seem suitable for running bipeds. The tails of some dinosaurs seem to have served as weapons.

Principal Sources

Chapter 5, like chapter 4, is based largely on Alexander (1985), with some of the experiments modified. The data about blood pressure are from Hohnke (1973). The suggestion that some sauropods may have reared up on their hind legs was made by Bakker (1978). The elastin ligaments in the necks of birds were investigated by Bennett and Alexander (1987). Galton (1970) described the stiff tails of duck-billed dinosaurs. Alexander and Vernon (1975) discussed the oscillations of the tails of kangaroos.

Alexander, R. McN. 1985. Mechanics of posture and gait of some large dinosaurs. *Zoological Journal of the Linnean Society* 83:1–25.

Alexander, R. McN. and A. Vernon, 1975. The mechanics of hopping by kangaroos (Macropodidae). *Journal of Zoology* 177:265–303.

Bakker, R. L. 1978. Dinosaur feeding behavior and the origin of flowering plants. *Nature* 274:661–663.

Bennett, M. B. and R. McN. Alexander, 1987. Properties and function of extensible ligaments in the necks of turkeys (*Meleagris gallopavo*) and other birds. *Journal of Zoology* 212:275–281.

Carpenter, D. 1982. Skeletal and dermal armor reconstruction of *Euoplocephalus tutus. Canadian Journal of Earth Sciences* 19:689–697.

Galton, P. M. 1970. The posture of hadrosaurian dinosaurs. *Journal of Paleontology* 44:464–473.

Gregory, W. K. 1951. *Evolution Emerging.* New York: Macmillan.

Hohnke, L. A. 1973. Haemodynamics in the sauropods. *Nature* 244:309–310.

Norman, D. B. 1980. On the ornithischian dinosaur *Iguanodon bernissartensis* of Bernissart (Belgium). *Mémoires de l'Institut Royal des Sciences Naturelles de Belgique* 178:1–103.

VI

Fighting and Singing Dinosaurs

THE TAILS mentioned at the end of chapter 5 are not the only weapons that dinosaurs had. I will now tell about horns and reinforced heads, and about how they may have been used for fighting. I also speculate about how dinosaurs may have approached rivals, and about the noises they may have made.

It seems likely that male dinosaurs fought rival males of the same species, much as stags fight with their antlers and male antelope fight with their horns. It may seem odd that animals fight, getting hurt and possibly killed. You may think that peaceful species should flourish and quarrelsome ones should destroy themselves, and that evolution by natural selection should make all species peaceful.

To understand why it does not, you need to understand the principle of the survival of the fittest. This principle is not about the survival of species or even the survival of individual animals: it is about the survival of genes, the basic units of heredity. Genes are complex molecules that carry information. Some carry patterns for making other molecules. Other genes or groups of genes carry instructions for making parts of the body, or for patterns of behavior. Almost every animal (there are some exceptions) inherits half its genes from its mother and half from its father. In turn, it passes on copies of just half of its genes in each egg or spermatozoon that it makes. Thus genes are transmitted from generation to generation. Errors of copying produce new genes from time to time, and the mixing of genes that happens in reproduction brings new combinations of genes together. Some genes or groups of genes make their possessors more successful in reproduction, and are more likely than others to be passed on to successive generations.

Some of them are obviously beneficial. For example, a gene that made its possessors immune to disease, or enabled them to run faster to escape predators, would make them better able to survive and reproduce.

It would be more likely than alternative genes, that did not give these advantages, to be passed on to successive generations. Once such a gene had appeared it would be likely to become commoner until, many generations later, all members of the population had it.

Other genes that are less obviously beneficial will also increase. Suppose that a gene made its male possessors very aggressive, quick to grab and copulate with every female they met and to chase other males away. Males with that gene would be apt to get involved in a lot of fights, to get hurt and to die young, but they might also get unusually many offspring. If they did, the gene would be favored by natural selection.

It is much more likely that males will fight for females than that females will fight for males. The reason is that females have to put a lot of energy and materials into making each egg or embryo, so they can produce only a limited number of offspring. Males, however, can produce enormous numbers of sperm. A female can get all her eggs fertilized by a single father but a male has plenty of sperm for many mates. If the numbers of males and females are about equal (as they are for most species) the females will easily get all the matings they can use but the males will have plenty of sperm to spare. The males could get more offspring (and pass more genes on to future generations) if they could copulate with more females and keep other males away from them.

Triceratops has horns that must surely have been weapons (figure 1.13). It must have looked rather like a huge rhinoceros with a peculiar arrangement of horns, but these are more like antelope than rhinoceros horns. Antelopes have spikes of bone covered with horn but rhinoceros horns are consolidated tufts of hair with no bony core. *Triceratops* horns are spikes of bone and were presumably covered with horn when the dinosaur was alive.

Many of the large, plant-eating mammals have horns of some kind. Antelopes, cattle, and sheep have true, permanent horns, and deer have antlers which they shed and re-grow annually. All these animals may sometimes use their horns to defend themselves against predators, but their usual defense is to run away. The principal use of their horns and antlers seems to be for fighting between males of the same species.

Red deer are a European species, closely related to the American wapiti. Their stags assemble harems, usually of about six females but sometimes of twenty, and defend them from other males. Only stags that have harems breed, but each of them may father many offspring. Genes that help stags to get and keep large harems must be favored by natural selection.

Stags that have no harem try to get one and those that have a harem already try to add to it. This leads to fighting between rival males. They run at each other and clash antlers. The antlers interlock (figure 6.1a) and the stags wrestle, each trying to push the other back or to throw it to the ground. A stag that manages to throw another may stab it with the points of its antlers, but the loser usually gives up and runs away without being thrown. Antelopes interlock horns and wrestle in the same way.

Many other species of deer defend harems, as do many antelopes, but other animals with different social systems fight in different circumstances. Bighorn sheep do not keep harems, but males seek out females that are in estrus and guard just one female at a time. They stay with her for one to three days, while she remains in estrus, copulating about once an hour to make sure that their sperm are in the right place at the right time, when the female ovulates. Other males try to capture estrus females from males that are guarding them, or to sneak a quick copulation while the guarding male is distracted. This leads to fighting. Sheep horns do not interlock, so there is no wrestling. Bighorns and other wild sheep fight by running at each other and colliding head-on, or by rearing up and clashing horns (figure 6.2).

If two *Triceratops* ran at each other with their heads down, their horns would interlock (figure 6.1b). It seems likely that males inter-

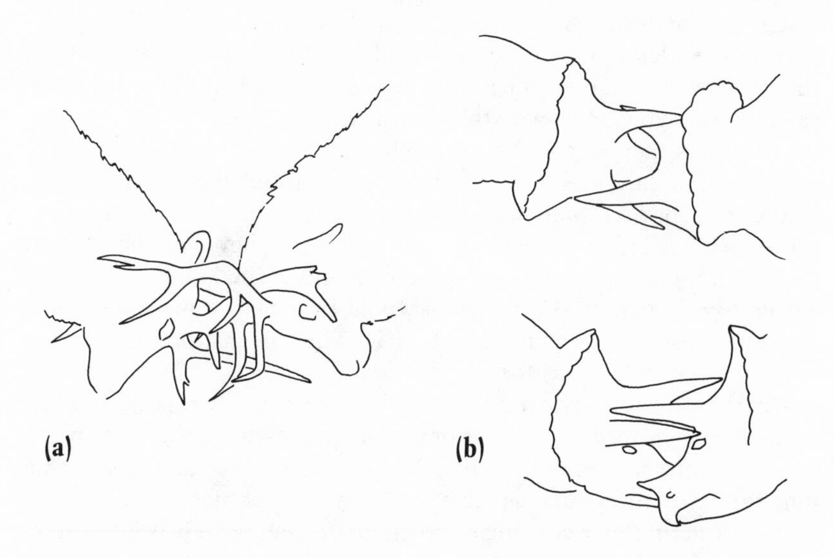

FIGURE 6.1. (a) Red deer stags with their antlers interlocked, drawn from a photograph in Clutton-Brock 1982; (b) *Triceratops* fighting with their horns interlocked, drawn from photographs of models sold by the Natural History Museum, London.

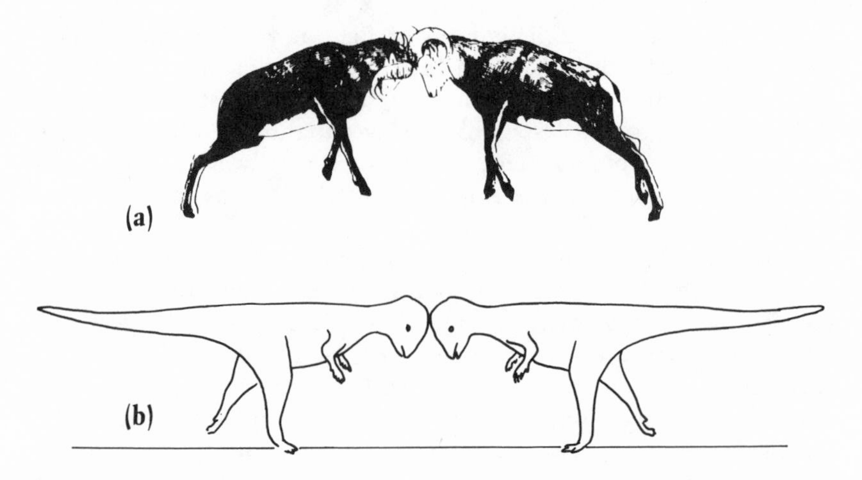

FIGURE 6.2. (a) Thinhorn rams fighting, from Geist 1971; (b) *Stegoceras* fighting.

locked horns and wrestled like stags and antelopes, but it seems just a little doubtful whether their horns were strong enough.

Figure 6.3 shows that *Triceratops* had rather small horns for its size, by antelope standards. Horn reach (figure 6.3a) means the straight-line distance from horn base to horn tip. The reach of a 6-tonne *Triceratops* is little different from that of an eland of one-tenth its mass, and far less than what the lines suggests 6-tonne antelope would have, if antelope grew so big. However, the shortness of the horns might not matter, provided they were strong enough.

Figure 6.3b shows the cross-sectional areas of the bony cores of horns, at the horn base. Female antelope have thinner horns than males of the same species, or no horns at all (but when they have horns they are generally about as long as those of males). The graph shows that *Triceratops* horns are thinner than would be expected on antelopes of the same mass, whether male or female. That means that *Triceratops* horns were rather weak, for the size of animal.

We need to be rather careful about the argument here, because short horns may not need to be as strong as long ones would have to be on the same animal. Remember the passage in chapter 4 where I said that long thin structures such as leg bones or horns are most easily broken by forces acting at right angles to them, which set up bending moments in them. The maximum stress at a distance x from a force R acting at right angles to the horn is Rx/Z, where Z is the quantity called the section modulus. A thin horn has a low section modulus but, if it is also short, it has low x as well as low Z. The stress set up in it by a force R may be no more than in a thicker, longer horn.

We might think that the shortness of *Triceratops* horns made up for their thinness, but for two things. First, wrestling antelopes engage the bases of their horns, not the tips, so x may not be proportional to horn length. Second, figure 6.3 shows little sign of a trade-off between thickness and length. The oryx has remarkably long horns that are also rather thin for the size of antelope and the wildebeest has short horns that are rather thick. There is no question of horns being so strong that differences of thickness do not matter: 3 percent of male wildebeest and 17 percent of male oryx, counted in wild populations in East Africa, had one or both horns broken. *Triceratops* horns seem rather weak, for so large an animal.

You might suppose that horns used mainly for fighting between males would be grown only by males. Antlers are indeed grown only by male deer, with one exception: both sexes of caribou have them. Horns, however, are grown by both sexes of many antelope species, as figure 6.3 shows, and also by both sexes of cattle and sheep. If ceratopians used their horns mainly for fighting between males you should not necessarily expect to find horns in males only, but you might reasonably expect the horns to be thicker in males than in females. The fossils show no clear signs of this. So far, scientists have been unable to decide which are the male skulls and which the female ones, for any of the horned ceratopian species. (They think they have worked it out for *Protoceratops,* which has no horns.)

When a male deer or antelope challenges another, they do not fight immediately. Each puts on a display which enables the other to gauge its strength. In the case of Red deer the first stage is a roaring match. Each stag roars loudly and repeatedly. As the contest continues, roars follow each other at shorter and shorter intervals. If the challenging stag finds he cannot manage as many roars per minute as his opponent, he may give up and go away: roaring rate seems to be used as an indicator of strength. If he does not give up at this stage the stags walk side by side for a while, looking at each other and judging each other's strength. Again, the challenger may give up, but otherwise the stags fight.

The elaborate ritual before fights seems to have evolved for a good reason. A belligerent stag whose genes made him fight for females on all possible occasions might be tremendously successful in reproduction if he happened to be strong enough to win every fight. More probably, he would often lose fights. He might get badly beaten up while still quite young and leave very few offspring.

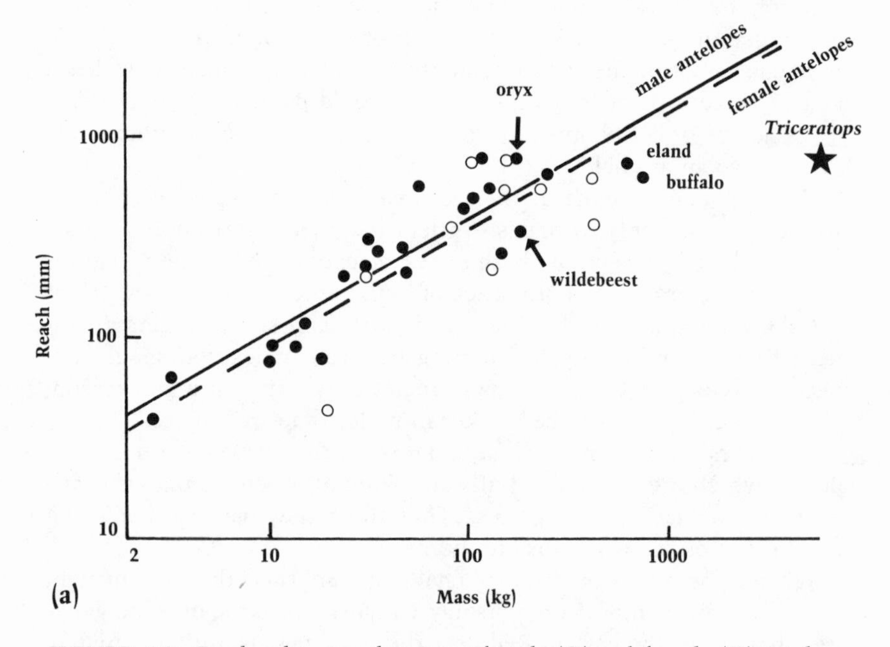

FIGURE 6.3. Graphs showing the sizes of male (●) and female (○) antelope and *Triceratops* horns. Graph (a) shows horn reach and (b) shows the cross-sectional area of the base of the horn, plotted in each case against body mass.

Now imagine a different set of genes that makes stags behave in a much more calculating way, that makes them fight only when they think they can win and run away when they are sure they would lose. Stags with these genes are likely to live longer and leave more offspring than stags that fight indiscriminately. The better they are at judging strength the more offspring they are likely to leave, but even if their judgment is not very good it should help. The set of genes is likely to be favored by natural selection. That is why Red deer have evolved the strength-assessing ritual. It seems likely that ceratopians did something rather similar.

Figure 6.4 shows the frill at the back of *Triceratops'* head, as well as the horns. This frill was formed by a strong sheet of bone, extending

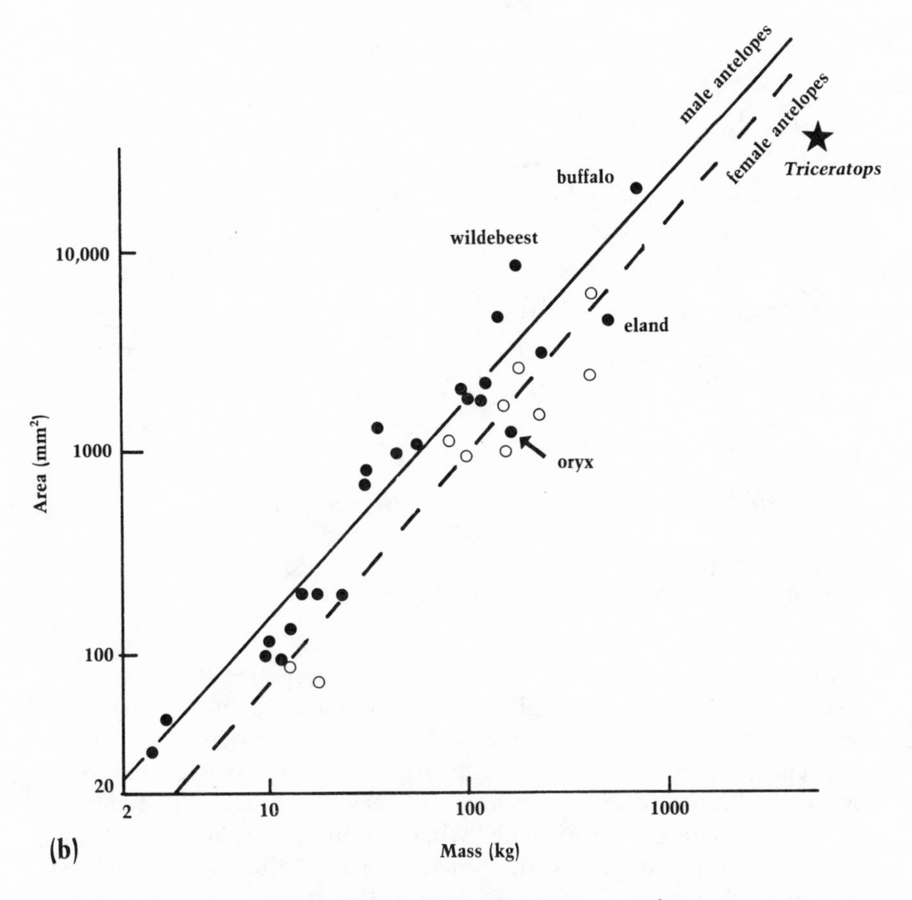

The antelope data are from Packer 1983. The *Triceratops* horn measurements are from Hatcher 1907, and the *Triceratops* mass is from chapter 2.

backward from the skull. There seem to have been big jaw muscles whose attachments extended onto the bony frill, but the frill seems too large to have evolved just for that. It must have been a useful shield, protecting the animal's neck from rivals' horns, but some other ceratopian frills were supported by bone only round the edge and would have been less good as shields unless their skin was very thick. The frill of *Styracosaurus* with its splendid row of spikes seems designed as much for show as for protection (figure 6.4).

Perhaps we can find a parallel among deer. Red deer have fairly practical-looking antlers: the spikes are weapons and the branches (especially the ones over the eyes) give protection by catching rivals' antlers. Moose, Fallow deer, and some others have palmated antlers, incorpo-

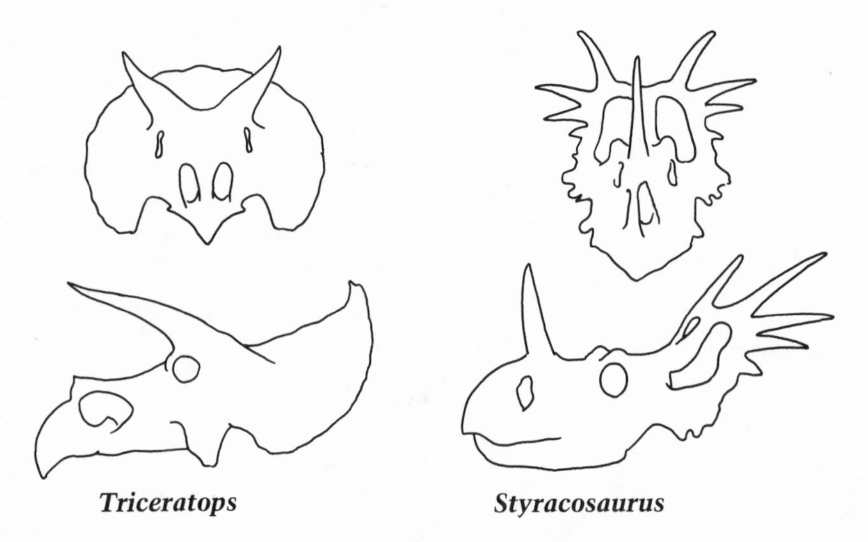

FIGURE 6.4. Skulls of *Triceratops* and *Styracosaurus*, drawn from illustrations in Hatcher 1907 and Brown and Schlaikjer 1937.

rating broad plates of bone that look most impressive but seem unnecessary for any mechanical function. They may help to attract females, like the long tails of male widowbirds which I discuss later in this chapter. Female Red deer mate with the stags that win them but Moose and Fallow deer have different social systems, and their females may have more choice of mate. Whatever the function of the bony plates on palmated antlers, the function of the frill of *Styracosaurus* was probably similar.

Now we will look at another probable weapon, the thick skull roof of the pachycephalosaurs (also called dome-headed dinosaurs). Figure 6.5 shows the skull of one of these animals (*Stegoceras*) and, for comparison, the skull of a similar-sized ornithopod. The huge bulge on the top of *Stegoceras'* skull is solid bone. Though the skull is only 28 centimeters long its roof is eight centimeters thick. *Stegoceras* is a small dinosaur, about two meters long, but some other pachycephalosaurs were much larger.

Their thick skull roofs look like battering rams. It seems likely that male pachycephalosaurs fought like Bighorn sheep, running at each other with heads down (figure 6.2). The thick skull roofs would take the impact, like the horns of the sheep. Some pachycephalosaur skulls have thinner roofs than others that seem to be the same species, and may be female. Similarly among sheep, ewes have smaller horns than rams.

We will think about the forces that might have acted when pachycephalosaurs collided. To simplify the calculation, imagine two identical dinosaurs running at equal speeds, colliding head-on. If everything is equal, neither is pushed back by the other: it is as if each collided with an absolutely rigid wall. If the dinosaur itself were rigid, it would be stopped instantly, but that would need an infinite force, which is impossible. The dinosaur would have to deform a little.

A running dinosaur has kinetic energy, which it loses when it stops. This energy equals half its mass times the square of its speed

$$\text{kinetic energy} = {}^1/_2 \text{ mass} \times (\text{speed})^2$$

The energy absorbed in the impact is the force multiplied by the deceleration distance

$$\text{energy absorbed} = \text{force} \times \text{deceleration distance.}$$

To stop the dinosaur, all the kinetic energy must be absorbed, so

$$^1/_2 \text{ mass} \times (\text{speed})^2 = \text{force} \times \text{deceleration distance}$$

$$\text{force} = \frac{\text{mass} \times (\text{speed})^2}{2 \times \text{deceleration distance}}$$

If the mass is measured in kilograms, the speed in meters per second and the distance in meters, this equation gives the force in newtons. The less rigid the dinosaur, the more it deforms, the bigger the deceler-

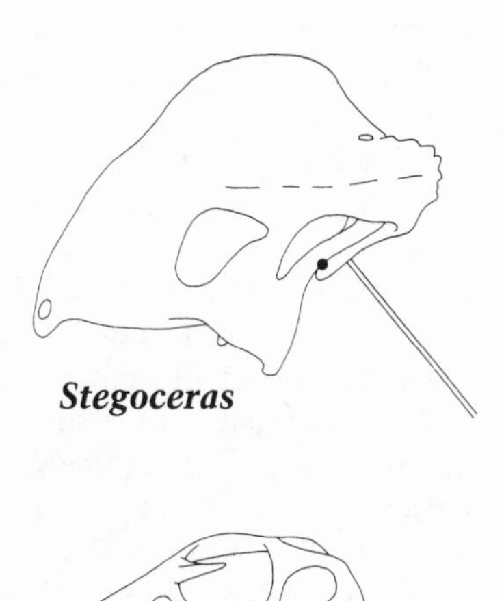

FIGURE 6.5. Skulls of *Stegoceras* (a pachycephalosaur) and *Hypsilophodon* (an ornithopod), from Sues 1978.

ation distance and so (from the equation) the smaller the force. The same is true of colliding cars, which is why crumple zones save lives. (Crumple zones are parts of a car body designed to crumple in an impact.)

A crumple zone is good for just one crash, but pachycephalosaurs presumably clashed heads often. What they would have needed, to moderate the forces, is some sort of elastic padding that would deform in the impact but spring back into shape afterward. The bulging skull roof may have been quite effective as padding. Sections cut through it show that the bone was slightly spongy, suggesting that it could have deformed quite a lot in an impact.

I want you to realize just how big the forces might have been, if the dinosaur were too rigid. Imagine a 20-kilogram dinosaur (a very rough estimate of the mass of *Stegoceras*) running at 3 meters per second. This would be a slow jogging speed for a human, and is well within the range of speeds estimated from dinosaur footprints in table 3.2. Suppose that dinosaur were in a collision that stopped it in a distance of one centimeter or 0.01 meter (it is hard to imagine the skull roof deforming more). The force would be $(20 \times 3^2)/(2 \times 0.01) = 9{,}000$ newtons, or 0.9 tonnes force. The skull roof of *Stegoceras* looks strong enough for that but I doubt whether the neck vertebrae could have stood it. (Unfortunately, no neck vertebrae have been found and we can only guess their strength from the sizes of trunk vertebrae.)

The double lines in figure 6.5 show the angles at which the neck vertebrae seem to have joined the skull. (This can be judged from the shape of the attachment area for the first vertebra, on the back of the skull.) The angle between the skull and the neck of *Stegoceras* means that, when the head was down in the butting position, the neck would have been more or less in line with the force. The trunk vertebrae of pachycephalosaurs interlock in a way that seems likely to have made the spine rather rigid, and there were bony tendons or ligaments in the back as well as the tail. Scientists have interpreted these signs as showing that pachycephalosaurs had very stiff backs and have pictured them colliding with their backs ramrod straight. I find that hard to believe because the forces (for any likely running speed) would be so large. They would be much smaller if the backbone buckled at impact; the skull would still be decelerated over a very short distance but most of the mass of the body would travel further forward, before being stopped. The neck could serve as a crumple zone, but one that could be straightened after the impact and used again.

Some of the hadrosaurs (duck-billed dinosaurs) also have strange projections from their skulls. *Anatosaurus* (figure 1.11) has a relatively plain skull but some other hadrosaurs with similar-shaped bodies have extraordinary crests (figure 6.6). *Corythosaurus'* crest is an upward-

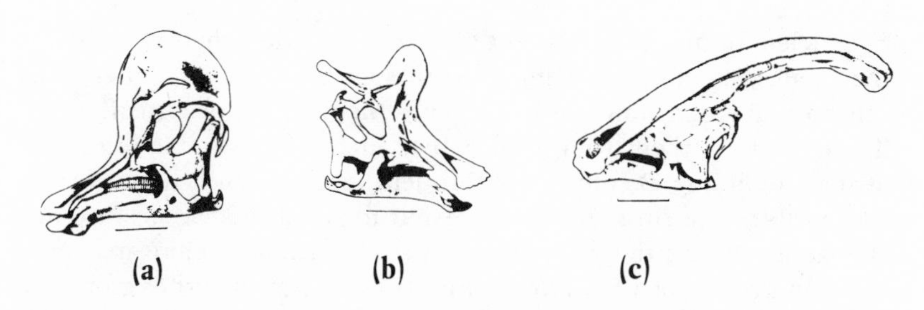

FIGURE 6.6. Skulls of crested hadrosaurs (a) *Corythosaurus;* (b) *Lambeosaurus;* and (c) *Parasaurolophus.* The scale lines are 20 cm long. From Weishampel 1981.

pointing semicircle, *Lambeosaurus'* crest has two branches and *Parasaurolophus'* crest is a backward-pointing rod. Other hadrosaur crests had other shapes. They do not look like weapons and they cannot have been very strong: some hadrosaurs had small solid crests but all the ones in figure 6.6 are hollow, with extensions of the nose cavity inside. Why did they evolve?

It seems likely that they evolved for the same reason as cocks' combs and peacocks' tails, as conspicuous ornaments. Among hadrosaurs that seem to belong to the same species there are some with large crests that were probably male and others with smaller crests that were probably female. Similarly, cocks have bigger combs (and tails) than hens, and peacocks have immensely longer, handsomer tails than peahens.

These structures seem to have evolved by a process called sexual selection: they evolved because females preferred highly ornamented males. The best evidence that this can happen comes from experiments with Long-tailed widowbirds in Kenya. These birds are about the size of American robins. The females are brown and inconspicuous but the males are black with extremely long (half meter) tails. They live in grassland, where the males establish territories averaging about a hectare (the area of a baseball outfield) and defend them against other males. They advertize themselves to females by flying low over their territories with their tails spread. Each female apparently selects a male, mates with him, and nests on his territory. Like Red deer stags, successful males mate with several females, but they do not compete for females by fighting each other. Instead, they compete to attract females.

Malte Andersson, a Swedish zoologist, did an experiment to test the hypothesis that longer tails are more attractive to females. He caught male widowbirds, altered the lengths of some of their tails, and released them again. He cut some of their tails down to one third of their original length and used the cut-off feathers to lengthen the tails of other males, by sticking them on with superglue. He left other birds with normal-length tails, either leaving them intact or cutting them off and sticking them back on again to check for possible effects of cutting and gluing. Before the experiment, each group had an average of 1.5 nests per territory (figure 6.7a), but many of the females had not yet mated. During the following month, males with elongated tails gained an average of 1.8 *additional* nests each, but the other groups of males gained only 0.5–1.0 additional nests per territory (figure 6.7b). The unnaturally elongated tails seem to have attracted females. This

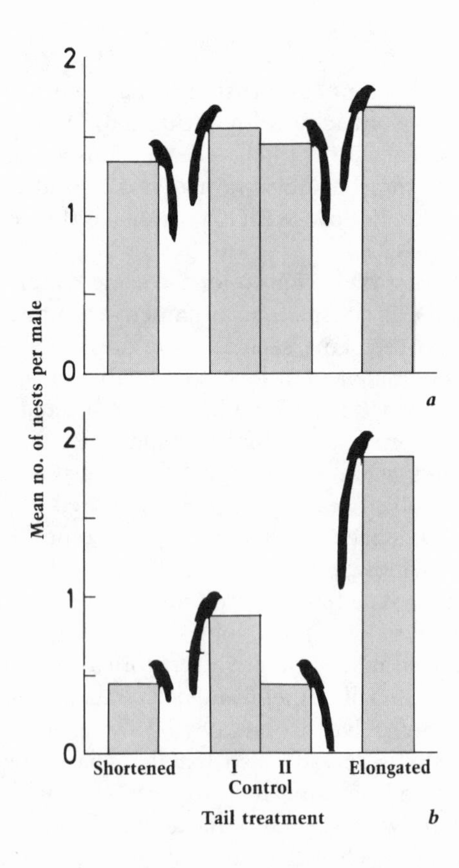

FIGURE 6.7. Widowbirds with long tails attract most females: (a) the mean numbers of nests per territory for four groups of males, before the experiment; (b) the mean numbers of extra nests built in the month after their tails had been shortened, elongated, cut and glued back on (control I) or left unaltered (control II). From Andersson 1982. Reprinted by permission. Copyright © 1982 Macmillan Magazines Ltd.

suggests that male widowbird tails evolved to their normal (already remarkable) length because the females liked them long. The normal length is probably a compromise between female preference and the inconvenience of an excessively long tail. Similarly, hadrosaurs probably evolved their crests because females liked them.

That leaves the question, why should females prefer anything so bizarre? Suppose you were a female hadrosaur who did not care for big crests, while most other females thought them marvelously sexy and desirable. You may think that a preference for big crests is just an irrational fad, but if you breed with a small-crested male your sons will probably have small crests and be unattractive to females, so you are unlikely to get many grandchildren. Once it has become fashionable to prefer big crests, genes that make females defy the fashion are likely to be eliminated.

Why should the fashion arise in the first place? I am going to tell quite a complicated story involving several of a female's relatives, and it will be easier to avoid confusion if she has a name (Dinah). Dinah has genes that make her prefer big crests, and she chooses a big-crested mate (Henry). Crest size is controlled by genes so it is likely that Henry's father also had a big crest. If he did, it is likely that Henry's mother preferred big crests (otherwise she would probably not have chosen him). This makes it likely that Henry has genes which, if he were female, would make him prefer big-crested males. Dinah's children are likely to inherit genes for preferring big crests not only from Dinah, but also from Henry. Futhermore, Dinah's mother probably preferred big crests so Dinah's father probably had one, and Henry's children will probably inherit genes for big crests from Dinah as well as from him. These effects are likely to make crests bigger and preferences stronger in successive generations. Mathematical analysis shows that this kind of evolution is likely to gather momentum and go to extremes.

One of the things that makes it seem likely that hadrosaur crests evolved by sexual selection is that their shapes are so different in different species. If they had a function other than fashion they would tend to have the same shape, the best shape for the job.

Figure 6.8 shows a *Parasaurolophus* crest sectioned to reveal the cavities inside. You can see how the nasal passages start on the snout, run all the way to the tip of the crest and double back to reach the mouth cavity. This is a long route from nostrils to mouth even in this short-crested (presumably female) skull. It is about two meters ($6^{1}/_{2}$ ft) in the long-crested (presumably male) *Parasaurolophus* shown in figure 6.6.

Tubes can be used to produce musical sounds, as in organs, flutes, and trumpets. They can be made to resonate at particular frequencies, emitting the regularly repeating waves of sound that we recognize as

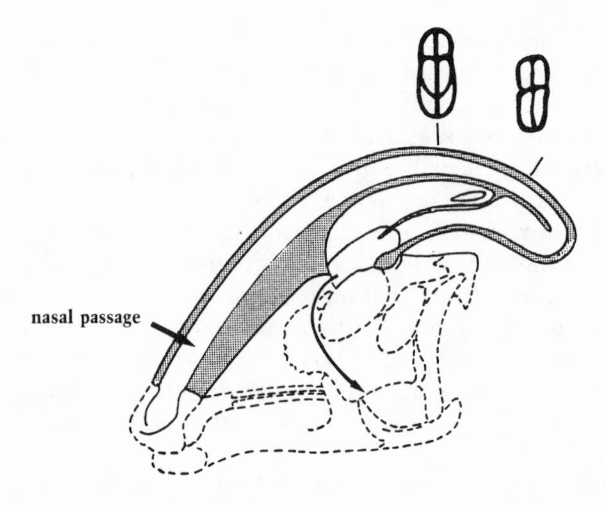

FIGURE 6.8. The crest of a *Parasaurolophus* skull sectioned to show the nasal cavities. This is probably a female skull, and it is not the same species as the *Parasaurolophus* in figure 6.6. From Hopson 1975.

musical. We also use the resonant properties of our nasal passages in voice production. It seems possible that *Parasaurolophus* used the long tubes in its crest to produce sounds. The males may have done this to attract females.

The resonant frequencies of air-filled tubes depend on their lengths. Long organ pipes give lower notes than short ones, and you get the lowest notes from a trombone by extending it to maximum length. I will try to explain why, using a different resonant system to introduce a basic principle.

Imagine a ball suspended on the end of a thin strand of rubber (figure 6.9). You can make it vibrate up and down by moving your hand, but there is one particular frequency of movement at which very small hand movements will sustain very large vibrations of the ball. That is the resonant frequency. It exists because of interaction between the mass of the ball and the elastic compliance of the rubber. "Compliance" may be an unfamiliar word. It means stretchiness, in this sense: if a force of one newton stretches the rubber by a centimeter, two newtons stretch it two centimeters, and so on, the compliance is one centimeter per newton. You can reduce the resonant frequency either by increasing the mass (using a heavier ball) or by increasing the compliance (using a longer or thinner piece of rubber).

The air-filled cavity in figure 6.9b resonates in essentially the same way. The vibrating mass is the air in the neck which vibrates in and out, compressing the air inside, which has elastic compliance. The air in the neck behaves like the ball and the air in the main cavity behaves like the strand of rubber but the distinction between the two lots of air is of course blurred. The distinction is even more blurred in figure 6.9c, which shows air vibrating in a tube that is closed at one end. The air near the mouth of the tube functions as a vibrating mass and the air near the closed end as an elastic compliance, but there is a gradation between the two. However, it is clear that a long tube contains a greater mass of air than a shorter tube of the same diameter. It also has higher compliance, because a long column of air is squeezed up more, by a given force on its end, than a short column would be. The air in a longer tube has more mass and compliance, so its resonant frequency is lower.

Figure 6.9d shows what happens in a tube open at both ends. The air at the ends vibrates in and out, compressing the air in the middle and allowing it to expand again. The resonant frequency is the same as for a tube with one closed end, of half the length. The air in a clarinet vibrates as in figure 6.9c but the air in a flute vibrates as in figure 6.9d because of the big hole at the blowing end.

The hollow crest of *Parasaurolophus* would probably have resonated as a tube open at both ends. The resonant frequency of such a tube is given approximately by a simple rule:

$$\text{frequency in cycles per second} = 170 \div \text{length in meters}$$

The length of the nasal passages in the crest of the *Parasaurolophus* in figure 6.6c is two meters, and the U-bend in them makes no differ-

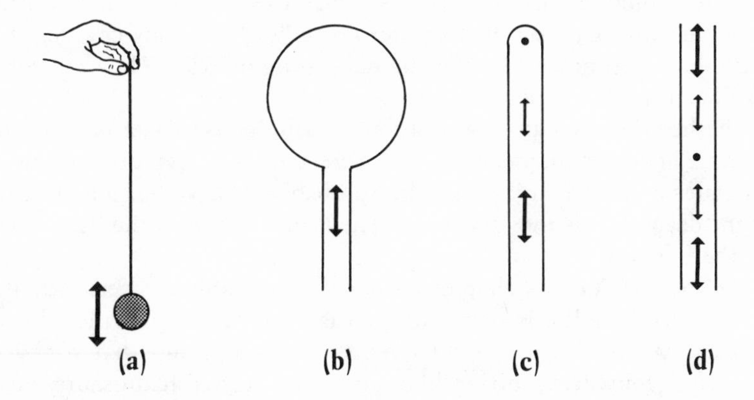

FIGURE 6.9. Diagrams of resonant systems: (a) a ball suspended by a strand of rubber; (b) an air-filled cavity; (c) a tube with one end closed; and (d) a tube open at both ends.

ence to the calculations. Their resonant frequency must have been about 170/2 = 85 cycles per second, close to the note

Females with shorter crests (figure 6.8) would have had higher-pitched voices. The equation makes it possible to estimate the pitch of the dinosaur's voice but tells us nothing about the tone, whether it was a pure single-frequency sound like a note from a French horn or enriched by many harmonics like a note from a bassoon.

A big dinosaur should have a deafening roar, or so you may think. However, there is a problem. You need a big loudspeaker to play a low note loudly, and trumpets would not be so loud if they were not flared out at the ends. The general rule is that to be most effective, loudspeakers and the mouths of trumpets should have diameters of at least one-sixth of the wavelength of the sound. The wavelength is the speed of sound (330 meters per second) divided by the frequency, so it is large for low frequency sounds. For a frequency of 85 cycles per second it is four meters, so an ideal loudspeaker for that pitch would have a diameter of almost 70 centimeters (28 in). *Parasaurolophus* should ideally have had nostrils flaring out like the mouths of trumpets 70 centimeters in diameter. They should have been at least as big as the mouths of tubas designed to play similar low notes. There is no sign of their being like that, so the dinosaur may not have been terribly noisy (if it sang at all): it may have been more like a bassoon (which plays low notes rather quietly) than a tuba.

Parasaurolophus had a long crest with a tube inside but the other hadrosaurs in figure 6.6 had shorter crests with bulbous cavities that would have resonated like the chamber in figure 6.9b. They too could have been used to produce sound.

Whether the resonators were tubular or bulbous, bigger ones would resonate at lower frequencies: larger crests would give deeper voices. If females preferred large crests they probably also evolved a preference for the deeper notes they produced. The hadrosaurs with the bass voices got the females.

The message of this chapter is that in many animal species, males compete for females. It seems likely that male ceratopians fought for females by wrestling with their horns like stags, and that pachycephalosaurs fought by butting like rams. The crested hadrosaurs, however, probably completed more peaceably, letting the females choose them. The females had evolved a preference for large crests and deep voices, and chose males partly for those qualities.

Principal Sources

Clutton-Brock (1982) tells about stags fighting and Geist (1971) tells about rams. The dimensions of antelope horns in figure 6.3 come from Packer (1983) and those of *Triceratops* from Hatcher (1907). Farlow and Dodson (1975) discussed the horns and frills of ceratopians and Sues (1978) described the skull roof of pachycephalosaurs. Andersson (1982) did the experiments on widowbird tails and Dawkins (1986) has written a good account of sexual selection. Hopson (1975) and Weishampel (1981) described and discussed the crests of hadrosaurs.

Andersson, M. 1982. Female choice selects for extreme tail length in a widowbird. *Nature* 299:818–20.

Brown, B. and E. M. Schlaikjer, 1937. The skeleton of *Styracosaurus* with the description of a new species. *American Museum Novitates* 955:1–12.

Clutton-Brock, T. H. 1982. The functions of antlers. *Behaviour* 79:108–125.

Dawkins, R. 1986. *The Blind Watchmaker.* London: Longman.

Farlow, J. O. and P. Dodson, 1975. The behavioural significance of the frill and horn morphology in ceratopsian dinosaurs. *Evolution* 29:353–361.

Geist, V. 1971. *Mountain Sheep.* Chicago: University of Chicago Press.

Hatcher, J. B. 1907. The Ceratopsia. *Monographs of the U.S. Geological Survey* 49:1–300.

Hopson, J. A. 1975. The evolution of cranial display structures in hadrosaurian dinosaurs. *Palaeobiology* 1:21–43.

Packer, C. 1983. Sexual dimorphism: The horns of African antelopes. *Science* 221:1191–1193.

Sues, H.-D. 1978. Functional morphology of the dome in pachycephalosaurid dinosaurs. *Neues Jahrbuch für Geologie und Paläontologie, Monatshefte* 1978(8):459–472.

Weishampel, D. B. 1981. The nasal cavity of lambeosaurine hadrosaurids. *Journal of Palaeontology* 55:1046–1057.

VII

Hot-Blooded Dinosaurs?

PEOPLE MAKE a big distinction between warm-blooded and cold-blooded animals. The warm-blooded ones are the mammals and birds, which feel warm and (if not too fierce) are nice to cuddle. Their bodies are wrapped in heat-insulating fur or feathers. The cold-blooded ones are snakes (and other reptiles), frogs (and other amphibians), fishes, and all the invertebrates. They usually feel cold when you touch them, and many of them have scaly or slimy skin. Were the dinosaurs warm-blooded or cold-blooded?

I want to rephrase the question. Scientists do not like talking about "warm-blooded" and "cold-blooded" animals because the "cold-blooded" ones are not necessarily colder than the "warm-blooded." Birds keep their bodies at 40–43°C and mammals at 36–40°C, in all climates. Reptiles, amphibians, and fishes have more varied temperatures that depend on their surroundings. Most fishes have temperatures almost exactly the same as the water they are living in, whether it be an arctic lake or a tropical swamp. Many reptiles, however, adjust their temperatures on sunny days by moving back and forth between sun and shade. If they stayed in the shade all day they would be rather cool and if they stayed in tropical sun all day they would get dangerously hot, but by moving in and out of the sun some manage to keep their temperatures in the mammal range for most of the day. Should we call them cold-blooded?

We scientists prefer to talk about ectotherms and endotherms. The ectotherms are the "cold-blooded" animals. If they control their body temperatures they do so mainly by using heat from the sun. The endotherms however depend mainly on heat produced within their bodies, by the chemical processes of metabolism. The best-known endotherms are birds and mammals but I will mention some mildly endothermic fish in chapter 9.

Metabolism supplies the energy needed to keep the body working. It happens in all living things, but particularly fast in endotherms. The most important processes of metabolism combine food with oxygen to give (mainly) carbon dioxide and water, and release energy. The necessary oxygen is obtained by breathing and the carbon dioxide is got rid of in the breath. The basic process is the same as burning, and the quantity of energy released is the same as if the food had been burnt to give the same end products. Some of the energy does useful jobs, powering muscles and driving the many energy-absorbing processes of life, but these processes are rather inefficient, and most of the energy is released as heat. The heat that warms our bodies is a waste product of metabolism.

The bodies of animals work by chemical processes, and chemical processes generally go faster at higher temperatures. High body temperatures enable animals to run faster than they otherwise could, to digest food faster, and (if food is abundant), to grow faster. For example, the lizard *Iguana* can run nearly three times as fast when its body temperature is 35°C, as when it is 20°C. Too high a temperature would be lethal but within limits it is good to be warm.

Metabolism goes fastest when animals are active but does not stop when they sit still. Even when resting they need energy to keep the heart beating and for many less obvious processes that are essential for life. When an animal is inactive its metabolism falls to a low, resting rate, which depends on the temperature. Ectotherms have higher resting metabolic rates at higher temperatures. Endotherms, however, increase their metabolic rates when they are put in *cold* places, to produce the heat needed to maintain their body temperatures.

The metabolic rates of mammals can generally be measured, by measuring how fast they use oxygen. This is done by putting the animal in a sealed container and analyzing samples of the air from time to time to find out how much of its oxygen has been used. (Obviously, the oxygen concentration must not be allowed to fall too much. Otherwise the animal's metabolism would be affected and it would eventually suffocate.) Whatever food is being metabolized, about 20 joules of energy are released for every cubic centimetre of oxygen used.

Figure 7.1 shows resting metabolic rates plotted against body mass, for lizards, birds, and mammals. It includes data for animals of a wide range of sizes, from sparrows to elephants. To show all these data clearly I have distorted the scales of the graph in the same way as I distorted the graph of bone circumference against body mass (figure 2.5): I have made the scales proportional to the logarithms of the quantities they represent so that the distance from 1 to 10 units is the same as from 10 to 100 units, or from 100 to 1,000 units.

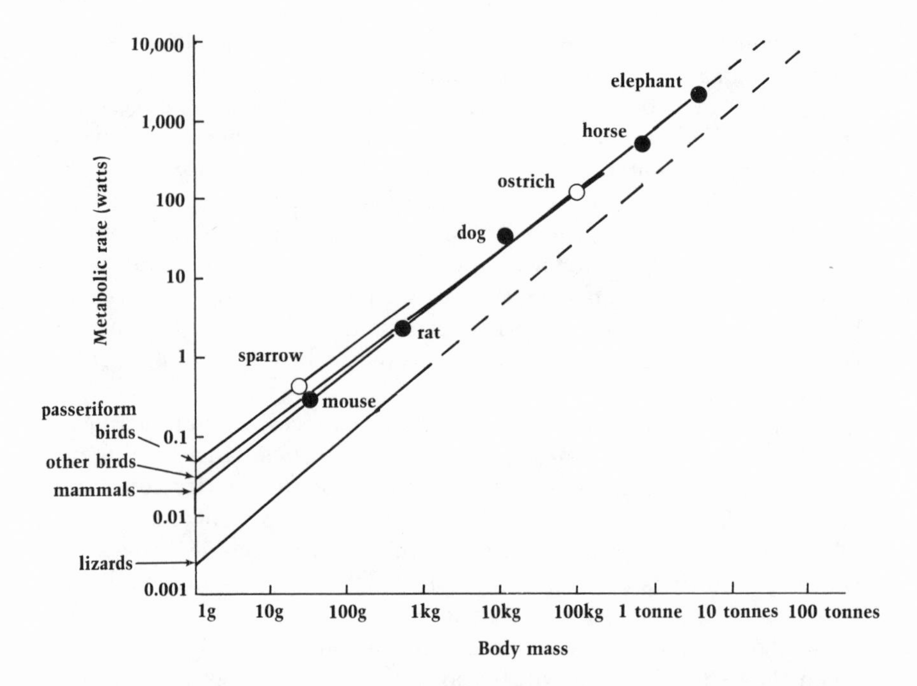

FIGURE 7.1. Graphs of metabolic rate against body mass for lizards, birds (○) and mammals (●). The passeriform birds (crows, finches, warblers, etc.) metabolize faster than other birds of equal mass. The line for lizards gives resting metabolic rates at 37°C. The data for birds and mammals are resting metabolic rates in the range of temperatures in which they are lowest. The lines are based on many more points than the few shown as examples. Sources of data are listed by Calder 1984.

The metabolic rates shown for lizards in figure 7.1 are resting rates at 37°C. The lizards could not have stood much higher temperatures, so these are near-maximum resting metabolic rates. The rates shown for mammals and birds, however, are minimum rates measured at comfortable environmental temperatures, at which the metabolism needed for other essential processes is enough to maintain the temperature of the body. The rates for mammals are nevertheless five to eight times as high as for lizards of the same mass, and some birds have even higher rates. Mammals and birds are enabled to be endothermic by their high metabolic rates. Even when they are resting at comfortable temperatures their metabolism produces heat fast enough to warm them quite a lot.

Dinosaurs are usually classed as reptiles. They are closely related to the crocodiles which, like other modern reptiles, are ectotherms. However, they are also closely related to the endothermic birds. Dr. Robert

Bakker argues that dinosaurs were endotherms. Some of his arguments are not (to me) very convincing. He points out that dinosaurs moved like mammals and birds, with their feet under the body (figure 3.5) and that some of the smaller ones were lightly built like ostriches. This suggests to him that dinosaurs, like ostriches, were fast-moving, fast-metabolizing endotherms. He also points out that the microscopic structure of dinosaur bone is like that of birds and mammals, which suggests to him that dinosaurs, like birds and mammals, were endotherms. (The microscopic structure can be seen in polished slices of fossil bone.)

These kinds of evidence are indirect. There is no clear reason why ostrich-like proportions or mammal-like bone structure should be found only in endotherms. However, Bakker presented a third argument with a much clearer logical basis. Endotherms metabolize much faster than ectotherms, so they have to eat a great deal more. An endothermic predator needs a bigger herd of prey to keep it supplied with food, than an ectotherm would do. If we can estimate the relative population sizes of carnivorous dinosaurs and their prey, we may discover whether the dinosaurs could have been endotherms. A population of endothermic carnivores would need a much bigger population of prey to support them, than would a similar-sized ectothermic population.

There are two difficulties about this line of argument. The first is that it is difficult to be sure just how big a population of prey is needed to feed a population of endothermic predators. Obviously reproduction and growth of the prey population must be enough to compensate for what the predators eat (otherwise the prey population will dwindle and the predators will starve) but it is hard to calculate how big a prey population that needs.

Ecologists have counted the large mammals in many nature reserves in Africa, India and America. They have estimated, for example, that each square kilometer of the Ngorongoro crater in Tanzania supports 10.4 tonnes of herbivores (antelopes and zebra) but only 96 kilograms of carnivores (lions and hyaenas): the mass of carnivores is only 1 percent of the mass of herbivores. In 29 other reserves the carnivore mass was always less than 2 percent of the herbivore mass. Bigger carnivore populations probably could not find enough food.

In contrast, in populations of ectothermic predators and prey, the predator mass may total 10 to 40 percent of the prey mass. This has been found, for example, for populations of spiders and ants feeding on insects. We would like to have data for communities in which both the principal predators and the principal prey were reptiles, but, unfortunately, there do not seem to be any modern communities like that. However, it does seem clear that, in a community of ecothermic dinosaurs, the predator mass could be far more than 2 percent of the prey mass.

The second difficulty about deciding whether dinosaurs could have been endotherms, by measuring the ratio of predators to prey, is that the ratio is very difficult to estimate for extinct animals. The best dinosaurs to try with (because there are lots of them) seem to be the ones that have been found in Canada in some late Cretaceous rocks, the Oldman Formation. A count of these dinosaurs gave 246 individual herbivores (mainly ornithopods) estimated to have an average mass of 5 tonnes, and 22 carnivores (tyrannosaurs) averaging 2 tonnes. This makes 1,230 tonnes of herbivores to 44 tonnes of carnivores. However, the commoner herbivores in the Oldman Formation are so very common that the paleontologists who collected there probably did not bother to collect the less good specimens. An informed guess about how many they left increases the estimated mass of herbivores to 2,110 tonnes. This makes the carnivore mass 2 percent of the herbivore mass.

If the dinosaurs had all been killed at once by a volcanic eruption or some other calamity, we could conclude that the carnivore population had had 2 percent of the mass of the herbivore population. There is no sign of such a calamity and we must suppose that the dinosaurs died naturally of various causes. In that case the numbers of fossils probably do not reflect the sizes of the living populations, but the rates at which they died. The herbivores probably had shorter lives than the carnivores, because they were liable to be attacked and eaten, so there is probably a bigger proportion of herbivores among the fossils than in the living population. The carnivore population probably had more than 2 percent of the mass of the herbivore population, which makes it doubtful whether they could have had mammal-like metabolism.

We might conclude that the dinosaurs had reptile-like metabolism, or at least that their metabolism was slower than would be expected for mammals of equal size. However, I would be wary of reaching any firm conclusion on this evidence. The conclusion could easily be changed if it turned out that paleontologists had discarded even more of the less-good herbivore fossils than has been supposed.

One argument that might be used against the suggestion that dinosaurs were endotherms is that they had neither fur nor feathers. Most mammals have fur and birds have feathers to retain the heat produced by metabolism, but the few fragments of dinosaur skin that have been preserved as impressions in rocks are bare and scaly like the skin of modern reptiles. However, these fragments are from the skin of large dinosaurs, and very large modern mammals (elephants, rhinoceros, hippopotamus) also have no fur.

Endothermy is a matter of degree. A lizard might be able to metabolize fast enough to keep it a degree or two warmer than its environment, but that would not make you want to call it an endotherm: it could warm itself far more by the ectothermic method of basking in the sun. We would not call dinosaurs endotherms unless their metabolism could keep them a good deal warmer than their environments.

Could they have done this, in spite of their having no fur? If they were warmer than their surroundings they would lose heat, and to keep a constant temperature they would have to produce heat fast enough by metabolism to replace the losses. I will try to calculate how warm they could have kept themselves both if they had reptile-like metabolism and if they had faster, mammal-like metabolism. I will base the calculations on the rates at which modern reptiles cool, when moved from a warm place to a cooler one, so we will need to know something about the physics of cooling.

To show how things cool, I baked a potato in its jacket in the oven. When it had cooked I pushed a thermometer into it and left it to cool on the kitchen table. Figure 7.2 shows how its temperature fell. Initially it was 93°C, which was 72° above room temperature. After 40 minutes it was about 36° above room temperature, after 80 minutes it was about 18° above room temperature and after 120 minutes it was only about 9° above room temperature: the temperature difference halved every 40 minutes. Hot objects in constant environments generally cool

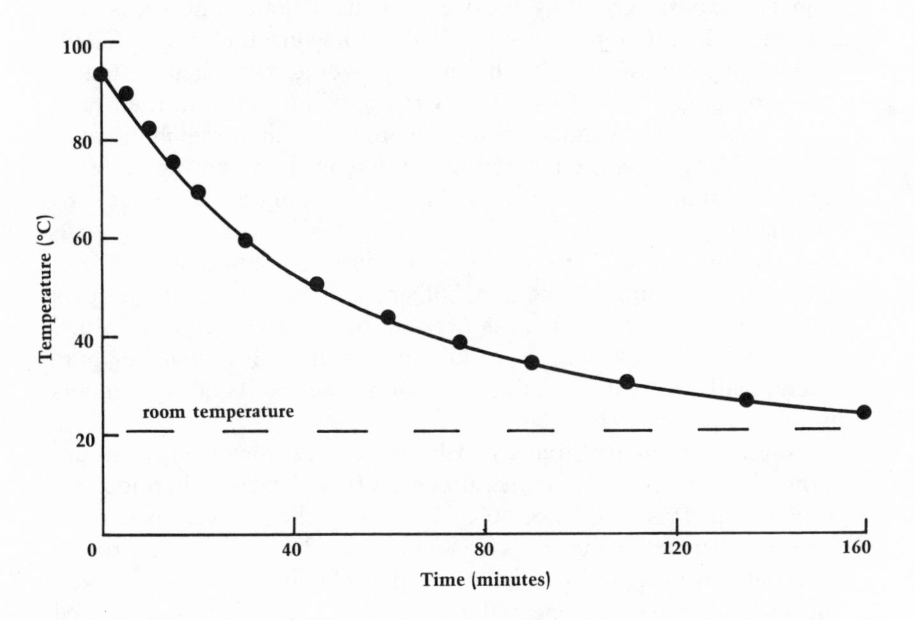

FIGURE 7.2. A graph of temperature against time for a cooling potato.

like this, taking equal times for each successive halving of the temperature difference (figure 7.3a). This is called exponential cooling.

Obviously, some things cool faster than others. We could use the half time (the time the temperature difference takes to halve) as a measure of the rate of cooling. Instead, I am going to do something that may seem capricious. I am going to use the time taken for the difference to fall to 0.37 of its initial value. This time is called the cooling time constant.

There is something very special about that number, 0.37. It is the reciprocal of the number called *e*, the base of natural logarithms, which is the second most useful number in the whole of mathematics. (The most useful is the one called *pi*). Figure 7.3b shows why it is useful to us here. The curve is exactly the same as in figure 7.3a. The temperature difference starts at D and reaches $0.37D$ after time T, so T is the time constant. Cooling starts fast and gradually gets slower, but the sloping broken line shows that if it had continued at its initial rate the temperature difference would have reached zero at time T. In other words the cooling rate was D/T when the temperature difference was D.

D/T is the rate of fall of temperature, but we want to calculate the rate of loss of heat. We do this by multiplying by the heat capacity C, the amount of heat energy that must be gained or lost to change the temperature of the object by one degree. Thus the rate of heat loss from an object, when its temperature is D above its surroundings, is CD/T.

We want to know how much dinosaurs would have been heated by their own metabolism. To keep body temperature constant, their rates of heat loss (CD/T) would have to be matched by their metabolic rates (R): $R = CD/T$. D is the temperature difference between the dinosaur and its environment, the quantity that we want to know. We get it by rearranging the equation $D = RT/C$.

We will use figure 7.1 to estimate R for different-sized dinosaurs with reptile-like or mammal-like metabolism. C is easy to estimate for a dinosaur of any particular mass because the heat capacity of animal tissue is about 3,500 joules per kilogram, a little less than for pure water. Finally, we will get values for T from the results of experiments on modern reptiles.

In these experiments, living reptiles were put somewhere warm until their body temperatures (measured by a tiny electrical thermometer in the gut) had risen to 35 or 40°C. They were then moved to a cooler place and their temperatures were recorded as they cooled. Their metabolic rates were too low to have a noticeable effect on their temperatures so they cooled exponentially, like the hot potato. Big ones cooled

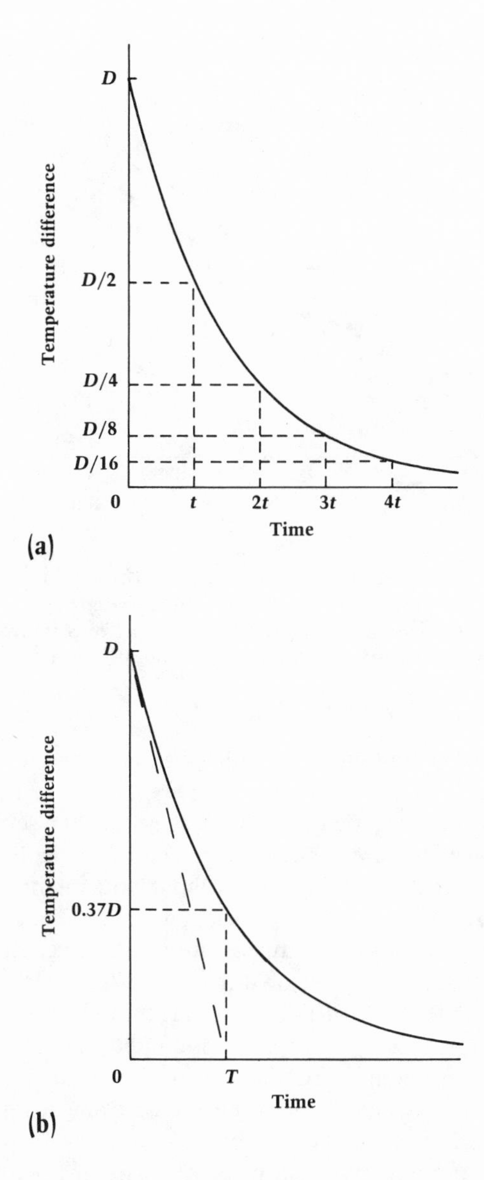

FIGURE 7.3. (a) A graph of temperature difference against time, showing the difference halving in successive equal intervals of time. D is the initial temperature difference and t is the time in which it halves. (b) The same graph, showing how the time constant T is defined.

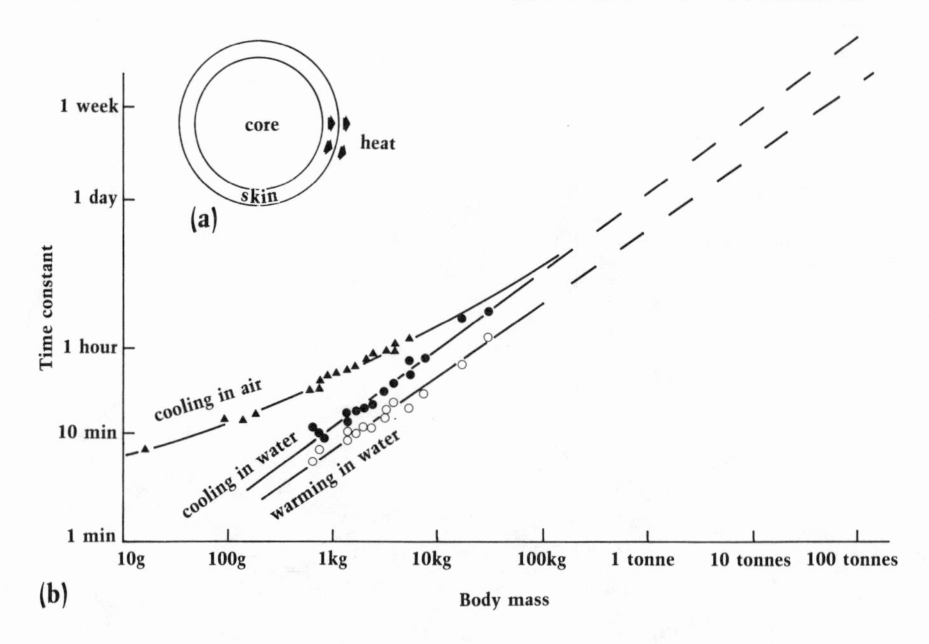

FIGURE 7.4. (a) A diagrammatic cross section through a reptile, illustrating the discussion of cooling. (b) A graph of time constant against body mass for various reptiles: ▲ cooling in air, in an air current of 3 metres per second; ● cooling in water; ○, warming in water. The data are from papers listed by Bell 1980.

more slowly than small ones just as big potatoes cool more slowly than small potatoes.

In some of the experiments, alligators and other semi-aquatic reptiles were cooled in water. Each reptile cooled faster in water than in air at the same temperature. This is what you might expect, but we need to understand why. The reptile's body can be thought of as a central core, kept at a uniform temperature by the circulating blood, and an outer heat-insulating layer of skin (figure 7.4a). Heat loss is a two-stage process. First the heat is conducted through the skin and then it leaves the skin surface by convection and radiation. Convection means the process of heat being carried away from the surface by moving air or water. These movements may be convection currents due to hot fluid rising or winds or water currents due to other causes. Convection works much more effectively in water than in air because water has a much higher heat capacity than an equal volume of air. This is why reptiles cool faster in water.

Figure 7.4b shows cooling time constants for a range of sizes of reptiles. The lines for cooling in air and in water converge, as body mass increases, suggesting that reptiles of 100 kilograms or more would cool almost as fast in air as in water. For them, heat loss would be limited almost entirely by the insulating effect of the skin. Theory predicts that the graphs should converge, when plotted on logarithmic scales as in figure 7.4b. The graph for cooling in air should be curved and the graph for cooling in water straight, which is how the graphs seem to be.

In other experiments reptiles were heated instead of being cooled: they were put in a cool place until their body temperatures had dropped to 15° to 20°C, and then moved to a warm one. Their temperatures were recorded as they warmed up and time constants were calculated. These time constants were shorter than the time constants for cooling, usually about half as long. I have included the time constants for warming in water in figure 7.4b but not those for warming in air, to avoid confusing the graph. If I had included the line for warming in air it would have run below and roughly parallel to the line for cooling in air.

Here is a likely explanation for the difference between cooling and warming. The reptiles probably let more blood flow to their skins when they were warming, reducing the effective thickness of the insulating layer of skin. They were being warmed from low temperatures to their preferred temperatures, so would want to warm as quickly as possible. Conversely, in the cooling experiments they were probably delaying cooling by restricting blood flow to the skin, making the insulating layer as thick as possible. People similarly adjust blood flow to their skin to help control body temperature: that is why we get flushed when we are overheated.

Dinosaur skin seems to have been like the skin of modern reptiles, so it seems reasonable to estimate time constants for dinosaurs by extending the graphs to higher body masses (figure 7.4b, broken lines). The difference between the warming and cooling graphs shows the likely scope for adjustment by controlling blood flow to the skin. The straightness of the lines for warming and cooling in water makes it seem fairly obvious how the extensions should go, but there is a danger of serious error because we have to extend the lines so far. Some dinosaurs were over 1000 times heavier than the biggest reptiles used in the experiments. It seems better to risk error than to give up in despair. The graph suggests that a 5-tonne *Iguanodon* would have had time constants of 2–4 days and a 50-tonne *Brachiosaurus* 1–3 weeks, depending on how much blood they let flow to their skins.

This graph gives us the time constants we need to estimate temperature differences between dinosaurs and their environments. In table 7.1 the "reptile-like" values assume that the dinosaur metabolizes at the rate indicated by figure 7.1 for a reptile of its mass, resting with its body at a temperature of 37°C. The "mammal-like" values assume metabolism at the rate indicated for a mammal resting in a comfortable warm environment in which the metabolism necessary for other purposes is enough to maintain body temperature. The range of temperature difference given in each case is from the value calculated from the warming time constant to the one calculated from the cooling time constant. It is the range that the animal could probably achieve by adjusting blood flow to the skin. Higher temperature differences would be possible if the animals increased their metabolic rates, as modern birds and mammals do in cold conditions. In experiments with Cardinal finches the birds doubled their resting metabolic rates when moved from an environment at 15°C to one at −15°C. Lower temperature differences than those shown in the table would also be possible if the animals panted to increase the amount of water lost by evaporation into their breath. The heat needed to evaporate the water would be lost from the body.

The table shows larger temperature differences for big dinosaurs than for small ones with the same type of metabolism, because bigger dinosaurs have longer time constants. It also (obviously) shows larger temperature differences for dinosaurs with fast, mammal-like metabolism than for ones with slow, reptile-like metabolism. It indicates that 50- and even 500-kilogram dinosaurs with reptile-like metabolism would be very little warmer than their surroundings: they would not be effective endotherms. Fifty-tonne dinosaurs with reptile-like metabolism would be quite good endotherms, up to 13° warmer than their surroundings. Moderate-sized dinosaurs with mammal-like metabolism would be endotherms, despite their lack of fur or feathers, and 50-tonne dinosaurs with mammal-like metabolism would be in danger of getting so hot that they cooked themselves, in most climates.

Small dinosaurs could not have been endotherms, even with mammal-like metabolism. Remember the baby *Psittacosaurus* in figure 1.2, smaller than a pigeon. It is much too small to have been an effective endotherm unless (which seems unlikely) it had fur or feathers. Some other baby dinosaurs were bigger, even when newly hatched, but it seems doubtful whether any were big enough to be endotherms from the start. Fossil dinosaur eggs are big, but they are not enormous. Even some thought to have been laid by a sauropod are only 25 centimetres long.

You would be wrong to assume that, because I have used scientific arguments, I have got everything right. I have made rough calculations involving extrapolation from the small reptiles used in experiments to the big dinosaurs. I will try to check whether the results are plausible by making some comparisons with modern animals.

Table 7.1 indicates that 50- and even 500-kilogram dinosaurs with reptile-like metabolism would be effectively ectothermic. This fits what we know of modern reptiles, all of which seem to be ectotherms. Even large crocodiles are ectotherms.

No modern reptiles have mammal-like metabolism but it seems reasonable to guess that mammals that have no fur have time constants about equal to those of similar-sized reptiles. Most of these mammals are large but the Naked mole rat (*Heterocephalus*) is small. It lives in tropical Africa in underground burrows in which the temperature remains very constant at about 30°C. Its body temperature is always close to burrow temperature. It is in effect an ectotherm, as table 7.1 suggests it would have to be. We ourselves are naked mammals with masses of about 50 (women) or 70 kilograms (men). Without clothes we are comfortable only in warm climates, which seems consistent with the calculation in the table that 50-kilogram dinosaurs with mammal-like

TABLE 7.1 Calculated temperature differences between dinosaurs and their environments. These are equilibrium values for dinosaurs resting in constant environments, taking no special steps to control body temperature. Cooling by evaporation of water is ignored.

Body mass	*50 kg*	*500 kg*	*5 tonnes*	*50 tonnes*
Metabolic rate (watts) for				
reptile-like metabolism	17	110	730	4,900
mammal-like metabolism	75	430	2500	14,000
Time constant (days) from				
cooling experiments	0.13	0.69	3.7	20
warming experiments	0.07	0.34	1.6	8
Heat capacity (megajoules per °C)	0.18	1.8	18	180
Temperature difference (°C) for				
reptile-like metabolism	0.2–0.4	0.7–1.4	2–4	6–13
mammal-like metabolism	3–5	7–15	20–50	60–140

The calculated temperature difference (°C) is:

$$\frac{\text{metabolic rate (watts)} \times \text{time constant (seconds)}}{\text{heat capacity (joules per °C)}}$$

metabolism could keep themselves only a few degrees warmer than their surroundings. Elephants seem to have trouble keeping cool in Africa, and seem comfortable enough even in winter in zoos in temperature countries, which seem consistent with the entry in the table for 5-tonne dinosaurs.

Those comparisons encourage me to think that the calculations may have been reasonably realistic, but another comparison worried me at first. The table says that a 50-tonne dinosaur with mammal-like metabolism would be at least 60° warmer than its surroundings unless it did something to get rid of excess heat. However, 50-tonne whales must always be less than 40° warmer than the water, even in polar seas. Remember that such large animals lose heat little faster in water than they would in air at the same temperature. If large whales are possible, should not equally large dinosaurs with mammal-like metabolism also be possible?

One fault of my argument so far is that I have almost ignored loss of heat by evaporation of water. I mentioned panting but have not considered the water loss that must occur all the time. The insides of lungs are damp, so water evaporates from them and gets lost in the breath. Skin is slightly permeable to water, so some water diffuses out through it and evaporates: crocodiles lose a lot of water this way, but desert lizards, with less permeable skins, lose only a little. I have written as if all heat loss from the bodies of reptiles happened by conduction through the skin followed by convection and radiation from the skin surface. We must remember that heat is also removed by evaporating water.

Suppose that a 50-tonne dinosaur were living in a warm climate, where the temperature of its surroundings was the same as its preferred body temperature. To keep its body at that temperature it would have to lose water fast enough to remove all the heat produced by its metabolism. The heat needed to evaporate one gram of water is 2,500 joules so if it had mammal-like metabolism, producing heat at the rate of 14,000 watts (joules per second; see table 7.1) it would have to evaporate about six grams of water per second, or about half a tonne (1 percent of its body mass) per day. A 5-tonne dinosaur with mammal-like metabolism, in similar circumstances, would have to lose about 90 kilograms of water (1.8 percent of body mass) per day. These rates do not seem impossible. A large (3.7 tonne) Indian elephant drank 140 kilograms of water per day and lost about 20 kilograms of it (0.5 percent of body mass) by evaporation. (The rest was lost in urine and feces.) This was when it was kept at a comfortable temperature, 20°C: in hot conditions it could presumably have allowed more to evaporate, and if necessary drunk more. Large dinosaurs with mammal-like metabolism need not have overheated, even in quite warm climates, if they allowed enough water to evaporate from their skin and into their breath.

I hope that argument has convinced you that you should not take the temperature differences in table 6.1 too literally. They could probably all be reduced to zero if the animals allowed enough water to evaporate from their bodies. However, they need not be far wrong for animals trying to conserve water: animals can, if necessary, lose a lot of water by evaporation and they must inevitably lose some (unless the atmosphere is very humid), but they need not lose much. The evaporation from the elephant at 20°C dissipated only 20 percent of the heat produced by its metabolism.

The ignoring of evaporation is not the only unrealistic thing about table 7.1. The table assumes resting metabolic rates, but animals do not rest all the time. Their metabolic rates are higher when they are active, and even at rest endotherms can increase their metabolic rates to maintain their body temperatures in cold conditions.

Despite these points the table has (I think) some value. Make any reasonable allowances for activity and water loss, and it will still show that only very big dinosaurs could have been effective endotherms, if they had reptile-like metabolism.

I have argued that even with mammal-like metabolism, big dinosaurs need not have overheated, if they allowed enough water to evaporate from their bodies. However, many hot environments are also dry, with water in short supply. It may be best to lose as much excess heat as possible by convection and radiation, keeping evaporation to a minimum. Elephants use their large ears as radiators and convectors. In hot conditions they dilate the blood vessels of their ears, and flap the ears to lose as much heat as possible from the blood passing through. It has been suggested that the plates on the backs of stegosaurs (figure 1.13) may also have served as cooling devices. Experiments with models in a wind tunnel seemed to confirm that the idea is feasible.

I have written so far as if an animal's surroundings had a clearly defined temperature. There is no problem for aquatic animals: for them the temperature of the water is the only external temperature that matters. A terrestrial animal, however, may be surrounded by air at one temperature, ground at another and foliage at a third. The atmosphere above it has different temperatures at different levels. The animal is affected by all these temperatures because it exchanges heat with the ground, foliage and sky by radiation, as well as exchanging heat with the immediately surrounding air by convection. Also, if the sun is shining its rays may have a profound effect on the animal's heat balance. There is, however, a temperature that we can think of as the effective temperature of the animal's surroundings. It is the temperature at which the animal's body would eventually settle if it did not metabolize or lose water by evaporation, and if conditions remained constant.

In practice, conditions do not remain constant. It is warmer during the day and colder at night. The body temperatures of modern ectotherms fluctuate accordingly. For example, an investigation of the lizard *Amphibolurus* in Australia showed that its body cooled to 25°C during summer nights but warmed to almost 40°C during the day. The temperatures of large dinosaurs must have fluctuated far less, because of their long time constants. They would hardly have started heating up during a hot day, before the cool night came. A 50-tonne *Brachiosaurus*, with estimated time constants of 8 to 20 days, must have had almost constant body temperature day and night. It may have got hotter in summer and cooler in winter, but its daily temperature fluctuations must have been slight. Even if it had reptile-like metabolism it could have maintained a high body temperature day and night, because of its huge size.

We started with the question, were dinosaurs ectotherms or endotherms? I explained the terms and showed that modern reptiles (which are ectotherms) have much slower metabolism than similar-sized birds and mammals (which are endotherms). I explained why dinosaurs have been thought to be endotherms. The most direct argument depended on the relative commonness of predatory dinosaurs and their prey.

I explained some of the basic physics of cooling and showed how rates of heating and cooling of modern reptiles could be used to estimate rates of heat loss from dinosaurs. The calculations seemed to show that small dinosaurs with reptile-like metabolic rates would be ectotherms (like modern reptiles) but that very large ones would be quite good endotherms. If dinosaurs had fast, mammal-like metabolism, very small ones (including hatchlings) would nevertheless be ectotherms unless they had fur or feathers. Moderate-sized ones would be endotherms and the largest would have been liable to overheat unless they lost a lot of water by evaporation from their skin and in their breath. Whether they had reptile-like or mammal-like metabolism, the body temperatures of large dinosaurs would have remained almost constant, day and night.

I have to admit after all this discussion that I do not know whether dinosaurs were ectotherms or endotherms, and whether they had mammal-like or reptile-like metabolic rates. Both questions remain to puzzle us. I would however like to present one thought that I find startling.

Suppose that the dinosaurs had reptile-like metabolism. Suppose also that plants grew as lushly as they do now, so that there was as much for herbivorous dinosaurs to eat as there is now for herbivorous mammals. If both those things were true, large dinosaurs may have been remarkably numerous. Figure 7.1 indicates that large dinosaur-sized reptiles would use energy at about the same rate as mammals of about one-fifth their mass. Therefore they need only as much food as mammals of one-fifth their mass. Think of the parts of East Africa where there are still large herds of mammals, including gazelles, wildebeest, zebra, buffalo, and elephant. Think of the population of dinosaurs that such a place could support if its vegetation was as plentiful in Mesozoic times. (The vegetation would have been different from modern vegetation but I see no reason why it should not have been as plentiful and have grown as fast.) For every 500-kilogram buffalo that lives now the vegetation could have supported a 2.5-tonne stegosaur, and for every 3-tonne elephant a 15-tonne *Diplodocus*. If dinosaurs had reptile-like metabolism they may have been as numerous as mammals of one fifth their mass in modern populations. The world may have seemed very full of dinosaurs.

Principal Sources

Bakker (1972) suggested that the dinosaurs were endotherms and stimulated a great deal of discussion which he has reviewed in a recent book (1986). Farlow (1976) made a more thorough investigation of the ratio of predators to prey in a community of large dinosaurs. Sources for my data on metabolic rates and on heating and cooling time constants can be found in Calder (1984) and Bell (1980) respectively. Spotila et al. (1973) took a different approach to dinosaur heat balance, making assumptions about the thickness of the skin instead of extrapolating from time constants of modern reptiles. Farlow et al. (1976) investigated the possibility that the plates of *Stegosaurus* may have been cooling fins. I have benefited greatly, in writing this chapter, from having seen some much more elaborate calculations about dinosaur heat balance, made by my student Jyrki Hokkanen.

Bakker, R. T. 1972. Anatomical and ecological evidence of endothermy in dinosaurs. *Nature* 238: 81–85.

Bakker, R. T. 1986. *The Dinosaur Heresies* New York: Morrow.

Bell, C. J. 1980. The scaling of the thermal inertia of lizards. *Journal of Experimental Biology* 86:79–85.

Benedict, F. G. 1936. *The Physiology of the Elephant.* Washington: Carnegie Institution.

Calder, W. A. 1984. *Size, Function, and Life History* Cambridge Mass.: Harvard University Press.

Farlow, J. O. 1976. A consideration of the trophic dynamics of a late Cretaceous large-dinosaur community (Oldham Formation). *Ecology* 57:841–857.

Farlow, J. O., C. V. Thompson, and D. E. Rosner. 1976. Plates of the dinosaur *Stegosaurus:* Forced convection heat loss fins? *Science* 192:1123–1125.

Spotila, J. R., P. W. Lommen, G. S. Bakken, and D. M. Gates. 1973. A mathematical model for body temperatures of large reptiles: Implications for dinosaur ecology. *American Naturalist* 107:391–404.

VIII

Flying Reptiles

THE PTEROSAURS lived at the same time as the dinosaurs. Both groups appeared late in the Triassic period and became extinct at the end of the Cretaceous. Pterosaurs were winged reptiles, with astonishingly long fingers that helped to support their wings (figure 8.1). Most fossils show only the skeleton, but a few show impressions of the wings and one has marks that seem to be impressions of fur. If pterosaurs were furry they were probably endotherms, with the fur serving as heat insulation. Flapping flight, as practised by birds and bats, is extremely energetic. Being endotherms may have enabled pterosaurs to get the necessary power output from their muscles.

Describing pterosaurs as furry animals with wings makes them sound pretty much like bats, but they looked quite different. Bat wings consist of a thin membrane stretched between the body, the arm and four extremely long fingers. Pterosaurs had only one finger, the huge "little" one, involved in supporting the wing, which was much more pointed than bat wings. Some of the wing impressions that have been found have striations on them, running in the same directions as the quills of the main wing feathers of birds (figure 8.1). If these are the remains of stiffening rods, as some palaeontologists think, the wings may have been relatively stiff structures like bird wings, not billowing membranes like the wings of bats and hang gliders.

Palaeontologists argue about the width of the wings. Figure 8.2a shows on the left the traditional reconstruction, a wide wing attached, as in bats, to the hind limbs as well as the fore. On the right it shows an alternative reconstruction that has been supported very persuasively by Dr. Kevin Padian of the University of California, Berkeley. This has narrow wings attached to the trunk but not to the hind legs. The fossils with wing impressions seem to show that the wings were indeed narrow.

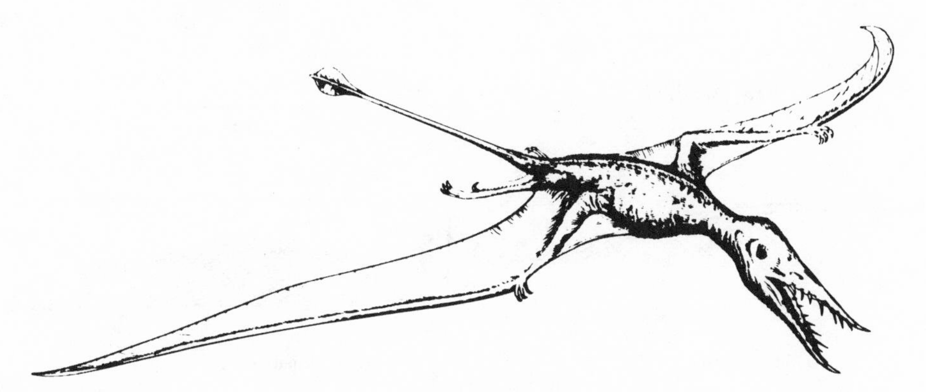

FIGURE 8.1. *Rhamphorhynchus,* a pterosaur. The wing span is one meter. From Wellnhofer 1975.

Figure 8.2b shows how supporters of the wide-wing hypothesis think pterosaurs would have looked on the ground. The hind legs stick out sideways like the legs of bats. The animal is imagined crawling awkwardly on all fours, hampered by the wing membrane attached to all four limbs. There is a famous set of fossil footprints that looks as if it was made by a pterosaur crawling like this but Padian has shown (surprizingly) that big lizards make very similar footprints. The marks that had been interpreted as impressions of the base of the long finger are imitated, in the lizard prints, by scratch made by a toe as the foot swings forward.

If the hind legs were not attached to the wings they need not have stuck out sideways. Padian thinks they were held like bird legs and that pterosaurs ran bipedally like birds (figure 8.2c). It ought to be possible to tell how the legs were held, from the structure of the hip joint, but this is difficult because most of the fossils have the pelvis crushed. The best fossils seem to show that the legs stuck out sideways, much as in lizards (figure 3.5a). This does not necessarily mean that pterosaurs crawled on all fours: some lizards get up on their hind legs to run.

Most pterosaurs were small, in the size range of common birds, but a few were impressively large. The biggest known from reasonably complete fossils is *Pteranodon,* which has a wing span of about 7 meters ($7\frac{1}{2}$ yds). I want you to realize how big this is. Try pacing out $7\frac{1}{2}$ yards. (If you do it indoors you will need a big room.) The biggest wingspan of any modern bird is only half as much—3.4 meters for the Wandering albatross. Even so, *Pteranodon* was not the biggest pterosaur. *Quetzalcoatlus* was even bigger with a span of (probably) 12 meters,

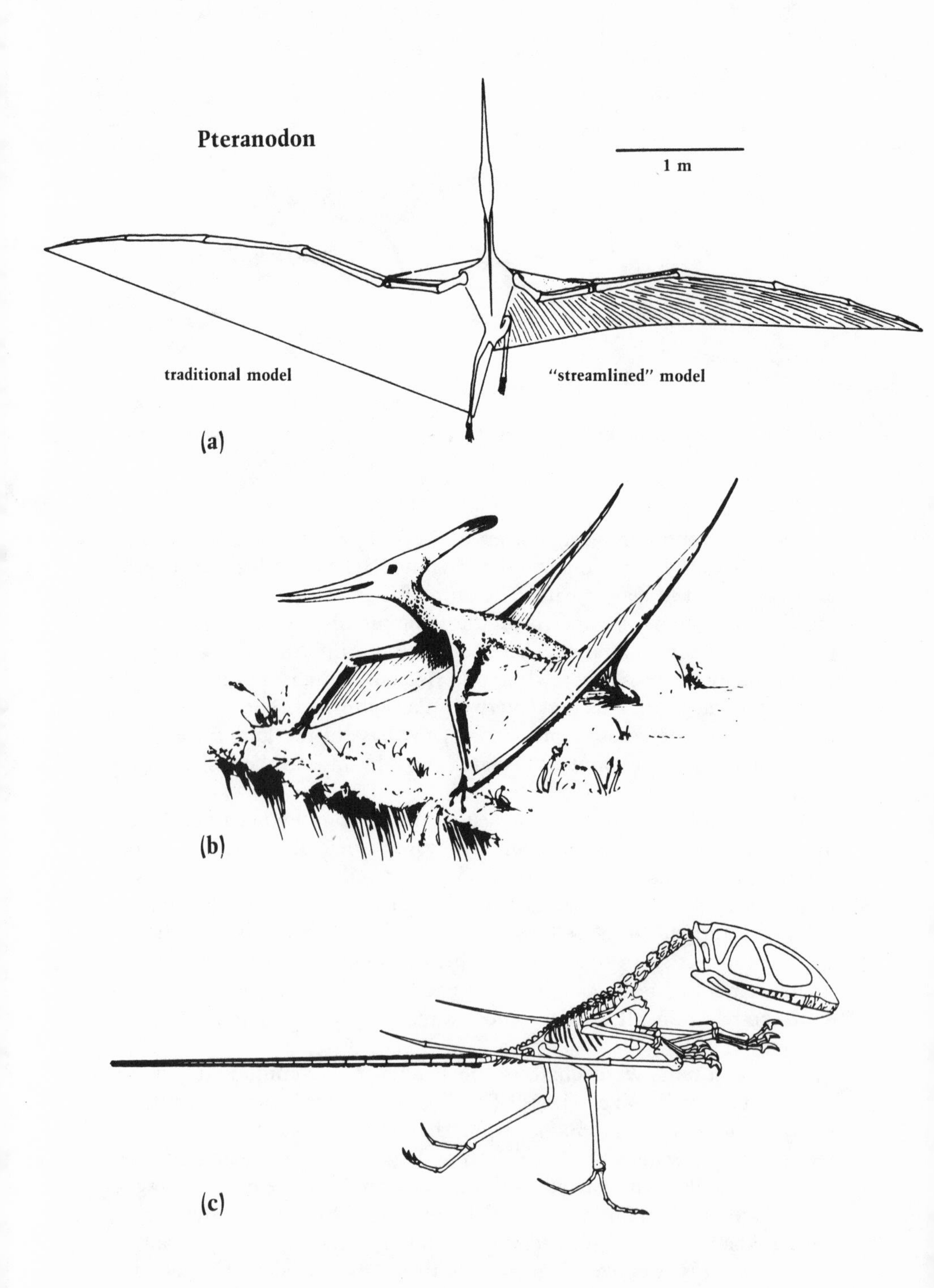
Pteranodon
1 m
traditional model
"streamlined" model
(a)
(b)
(c)

but the fossils found so far are very incomplete. *Pteranodon* is the biggest flying animal that we know much about, and most of the rest of this chapter is about it.

We cannot discuss how it lived without knowing something of aerodynamics, the basic science of flight. Figure 8.3a shows a section through a wing of an airplane, pterosaur or bird: it does not matter which. The air in front of the wing is stationary but the air behind is moving, set in motion by the passage of the wing. Notice that this air is moving forward and downward. It is moving forward because it has been dragged along by the passing wing. It is moving downward largely because the wing is tilted at an "angle of attack" to its direction of motion, but the arched ("cambered") shape of the wing section also helps to drive air downward. The air is being driven forward and downward so there must be backward and upward forces on the wing. The component that acts backward along the direction of motion is called drag and the component at right angles to the direction of motion is called lift. The lift is useful (it supports the weight of the flying aircraft or animal) but the drag is generally a nuisance. The lift can be made much larger than the drag by shaping the wing appropriately, and holding it at an appropriate angle of attack.

Figure 8.3b shows an airplane flying horizontally. Its wings must produce enough lift to balance its weight. Its propeller blows air backward, giving enough thrust to balance the drag on the wings plus the additional drag that acts on the fuselage. Flying animals have no propellers but flap their wings in such a way as to provide thrust as well as lift.

Figure 8.3c shows a glider. There is no propeller to give thurst but the forces are nevertheless balanced. The glider is moving on a downward slope, so the lift is tilted forward. The lift (acting upward and forward) and the drag (upward and backward) together balance the glider's weight.

Big wings can give more lift than small ones, at any particular speed. The same wing can give more lift when traveling fast, than when traveling slowly. Thus lift depends on wing area and on speed. It also depends on angle of attack: a bigger angle means more lift, up to a point, but if the angle becomes too large the lift falls again. The maximum

FIGURE 8.2. Contrasting views of the structure and posture of pterosaurs: (a) *Pteranodon,* showing a wide-winged reconstruction on the left and a narrow-winged one on the right. From Padian 1985; (b) *Pteranodon* reconstructed as a clumsy crawler. From Bramwell and Whitfield 1974; (c) *Dimorphodon* reconstructed as a nimble runner, From Padian 1983. The length of the head was about 1.8 meters (including the crest) in *Pteranodon* and 20 centimeters in *Dimorphodon.*

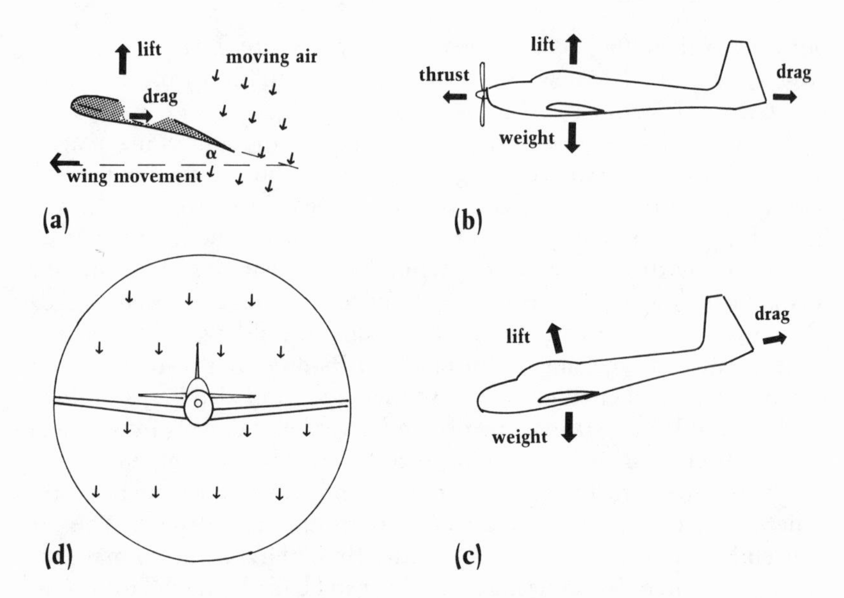

FIGURE 8.3. (a) A vertical section through a wing, showing the forces that the air exerts on it; (b) an airplane flying horizontally, with the forces that act on it; (c) a glider gliding with the forces that act on it; (d) a front view of an airplane, showing the air that is driven downward as the wings pass through it.

lift (L_{max}) that can be obtained by adjusting the angle of attack is proportional to the wing area A and the square of the speed v

$$L_{max} = \text{constant} \times Av^2$$

(Figure 8.4 shows how wing area is measured.) The constant depends on the shape of the wing, both in plan and in section, but is typically about 0.9 kilograms per cubic meter for well-designed wings. (For readers who already know about this sort of thing, this means that the maximum lift coefficient is about 1.5.) There are some ifs and buts here. What I have said would not be true for a very tiny wing or for a wing moving very slowly, but is about right for bird (or pterosaur) wings, moving in the range of speeds at which birds fly.

For an aircraft flying horizontally, the lift is fixed: it must equal the weight W of the body. This means that there is a minimum speed v_{min}. If the aircraft tries to fly slower than this, its wings cannot produce enough lift.

$$W = \text{constant} \times Av_{min}^2$$
$$v_{min} = \sqrt{(W/A)/\text{constant}}$$

W/A, the weight divided by the wing area, is called the wing loading, so if the constant is 0.9 kilograms per cubic meter

$$\text{minimum speed} = \sqrt{1.1 \times \text{wing loading}}.$$

(In this equation, speed is in meters per second and wing loading in newtons per square meter.) This equation tells us why jumbo jets need long runways. Imagine two airplanes of the same shape, one twice as long as the other. The big one has eight times the volume and so, probably, eight times the weight of the small one. However, its wings have only four times the area. Therefore it has twice the wing loading and 1.4 times the minimum speed (1.4 is the square root of two). Jumbo jets are not the same shape as small executive jets any more than swans are the same shape as sparrows, but the general conclusion holds: large aircraft cannot fly as slowly as small ones and so have to taxi to higher speeds to take off. *Pteranodon* was much bigger than modern flying animals. Would it have had trouble taking off?

Flying animals have an advantage over airplanes. They can flap their wings, moving them rapidly through the air although the body may be moving slowly. To take off, a small bird has only to jump into the air and start flapping its wings, but this needs a lot of power. Large birds cannot do this, but must get their bodies moving fast before they can take off. A large bird taking off from a cliff or branch can get up speed

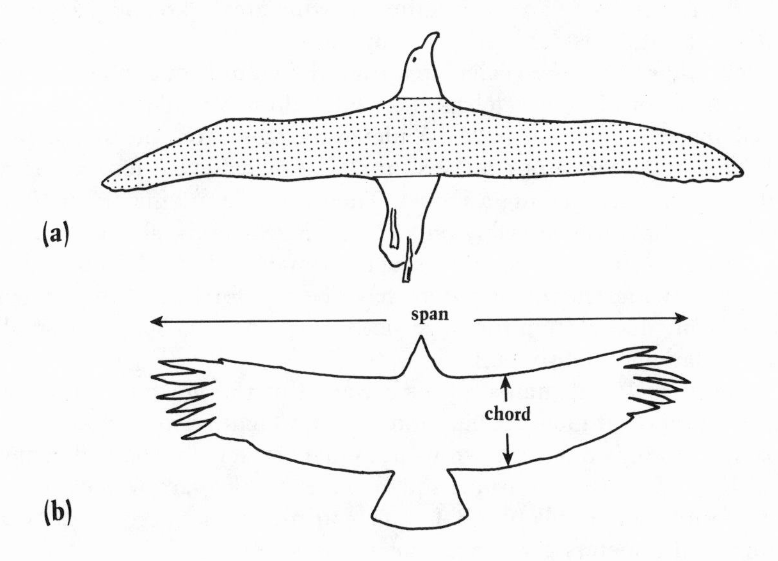

FIGURE 8.4. Outlines of (a) an albatross and (b) a condor, showing the meanings of some terms. Stipple in (a) shows how wing area is measured. Aspect ratio is (span/mean chord).

by diving from its perch, but to take off from level ground it must run like a taxiing airplane. It helps to run into the wind because the speed that matters is not the speed over the ground but the speed of the wings relative to the air.

The Kori bustard, found in the East African plains, is possibly the heaviest flying bird, with masses up to 16 kilograms. It seems to be a big effort for it to get airborne. It has to make a taxiing run, and often simply runs away when people approach, without bothering to take off. Vultures similarly run to take off, and swans and pelicans taking off from water run on the water surface to get up speed. All these groups include species that reach masses of 10 kilograms or more.

This suggests that it might have been difficult for the enormous *Pteranodon* to get airborne. Would it have had to run fast to take off, and if so could it take off at all? I cannot imagine the clumsy beast shown in figure 8.2b running, but *Pteranodon* may have moved more like the pterosaur in Padian's reconstruction (figure 8.2c), which looks quite fast.

It will help us to judge how difficult take-off would have been if we calculate *Pteranodon's* minimum flying speed. For this we need to know its wing loading, body weight divided by wing area.

We can measure the wing area from figure 8.2a, but should we use the wide-winged reconstruction (on the left) or the narrow-winged one (on the right)? They give very different wing areas, 4.6 and 2.5 square meters. I will consider both possibilities.

Body weight could be calculated from the volume of a model, as has been done for dinosaurs (chapter 2), but I do not think this has been tried. Instead, scientists have calculated the mass from the dimensions of bones and of drawings of the animal as they believe it looked in life, without actually making a model. They got a surprizing result: *Pteranodon*'s mass was probably only about 15 kilograms, about the same as a Kori bustard. This is only a rough estimate and may be quite badly wrong, but *Pteranodon* seems to have been a lot lighter than might have been guessed from the huge size of its wings. Its body was small, and remarkably lightly built.

A mass of 15 kilograms means a weight of 150 newtons. (Multiply mass by gravitational acceleration to get weight.) This gives a wing loading of $150/4.6 = 33$ newtons per square meters for the wide wings and $150/2.5 = 60$ newtons per square meter for the narrow ones. These give minimum speeds of $\sqrt{1.1 \times 33} = 6$ meters per second for wide wings and 8 meters per second for narrow ones.

These are remarkably low values, for such a large animal. The biggest albatrosses and vultures have smaller masses than *Pteranodon* but their wing loadings are about 150 and 100 newtons per square meter, respectively, giving minimum speeds of about 13 and 10 meters per second. It may have been easier for *Pteranodon* to take off than it is for these large birds. If the wind speed exceeded its minimum flying speed it would be really easy: it would only have to face into the wind and spread its wings, and up it would go. Winds of the necessary speed, 6 to 8 meters per second, are described by sailors as moderate breezes, and produce small waves with fairly frequent "white horses." Albatrosses take off by facing the wind and spreading their wings, but they need stronger winds because their minimum flying speeds are higher.

Pteranodon's low wing loading was partly due to the astonishing lightness of its big wing bones. These were air-filled tubes, with walls only about a millimeter thick (figure 8.5). Tubular structure is a good way of getting strength with lightness, when bending moments have to be resisted, because a tube has a bigger section modulus than a solid rod with the same amount of material in its cross section. (Section modulus was explained in chapter 4.) This is why bicycle frames and scaffolding are made of tubes. Most long bones, (including the ones in our own bodies) are tubular, but most of them are filled with marrow and so cannot be particularly light. Many bird bones are thin-walled air-filled tubes but none have such remarkably thin walls as the wing bones of *Pteranodon*. We are pretty sure that *Pteranodon*'s bones were air-filled because they have holes through their walls like the holes that connect the air cavities of bird bones to the lungs.

How did *Pteranodon* fly once it had taken off? Most large birds spend a lot of time gliding, using rising currents of air to keep themselves airborne. This is called soaring, and it seems likely that *Pteranodon* also soared. There are two principal soaring techniques used by different groups of birds. They are also used by glider pilots.

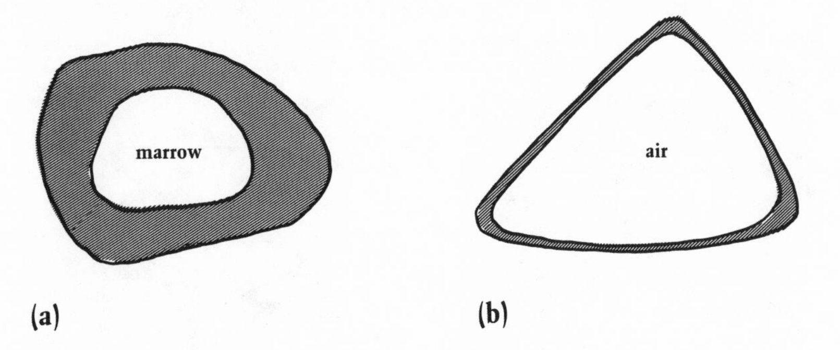

FIGURE 8.5. Sections of a typical mammal bone (a camel tibia) and of the first phalanx of the long finger of *Pteranodon*. Modified from Currey and Alexander 1985.

One of these techniques is called slope soaring. When a wind blows against a hillside or against the side of a wave it is deflected upward, making soaring possible. Figure 8.6a shows a bird soaring along the windy side of a wave. It is gliding, not flapping its wings, so it must be sinking relative to the air, but the air is rising. If the air is rising fast enough the bird can travel horizontally, keeping at the same height above the water. Albatrosses and petrels often soar along waves (but albatross also use another soaring technique that involves swooping up and down). Gulls similarly soar along hillsides and cliffs.

To slope soar without being blown downwind, a bird must glide at least as fast as the wind. If its air speed (its speed relative to the air) equals the wind speed, it can face directly into the wind and remain stationary relative to the ground. To travel at right angles to the wind it must glide faster than the wind, obliquely into the wind, as shown in figure 8.6b. (The ground speed, in this diagram, means the speed of the bird relative to the ground.) Slope soarers have to be able to glide fast.

The second important soaring technique uses thermals, which are columns of rising hot air. The sun shining on the ground heats it up, the hot ground heats the air immediately above it and the air over the hottest patches of ground rises as thermals. A bird gliding in a thermal

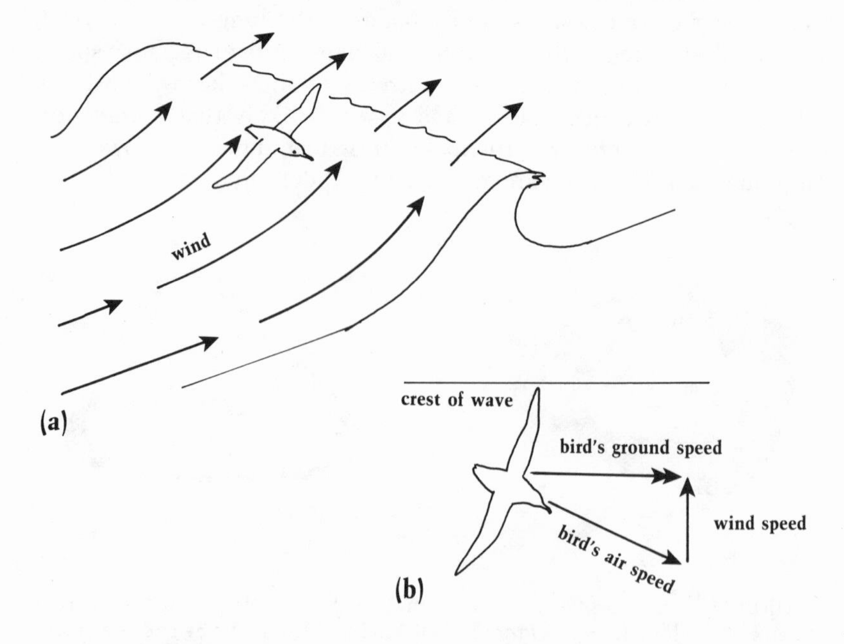

FIGURE 8.6. Diagrams of slope soaring.

gains height if its rate of sinking, relative to the air, is less than the rate at which the air is rising. Thermal soaring is done by circling for a while in a thermal, gaining height, and then gliding to the next thermal (figure 8.7a). Thermals are often easy to find because cumulus clouds (the fluffy kind) form at the top of them. Glider pilots find thermals by looking for the clouds, and birds presumably do the same.

Vultures soar all day in thermals, looking for carcasses to feed on. Some also travel long distances between nests and feeding sites by thermal soaring. Storks migrate between Europe and Africa largely by thermal soaring, making large detours to avoid the Mediterranean Sea, which has no thermals.

However, thermals do form over the sea in the parts of the tropics and subtropics where the trade winds blow. These thermals are produced by a different mechanism from the ones over land. The trade winds blow constantly and fairly gently from the northeast in the northern hemisphere and from the southeast in the southern hemisphere, carrying cool air toward the equator. This air is heated by the warm sea, and thermals form. Frigate birds soar in these thermals.

Thermal soaring over land or sea requires the ability to glide in small circles, because many thermals are only a few tens of meters across.

Slope soarers have to be able to glide fast and thermal soarers have to be able to glide in small circles. We may get clues about *Pteranodon*'s flying habits by asking which of these things it could do well.

Every glider glides well only in a limited range of speeds. If it glides slowly, near its minimum speed, it inevitably loses height rather rapidly. If it glides fast it again loses height rapidly. However, there is an intermediate range of speed at which low sinking speeds (relative to the air) are possible. I have already shown that the minimum air speed is proportional to the square root of wing loading. The air speed at which the sinking speed has its lowest value is also proportional to the square root of wing loading. This means that low wing loading is best for slow gliding and high wing loading for fast gliding. Slope soarers, which need to glide fast, should have high wing loading.

To glide in a circle, a bird needs a centripetal force pulling toward the center of the circle. (Similarly, a stone whirled on the end of a string is prevented by tension in the string from flying off at a tangent.) Birds get the necessary centripetal force by banking (figure 8.7b), so that the lift on their wings pulls inward (providing centripetal force) as well as upward (balancing body weight). Plainly, the lift must be bigger than the required centripetal force which is mv^2/r for a glider of mass m moving at speed v in a circle of radius r:

$$\text{Lift is greater than } mv^2/r.$$

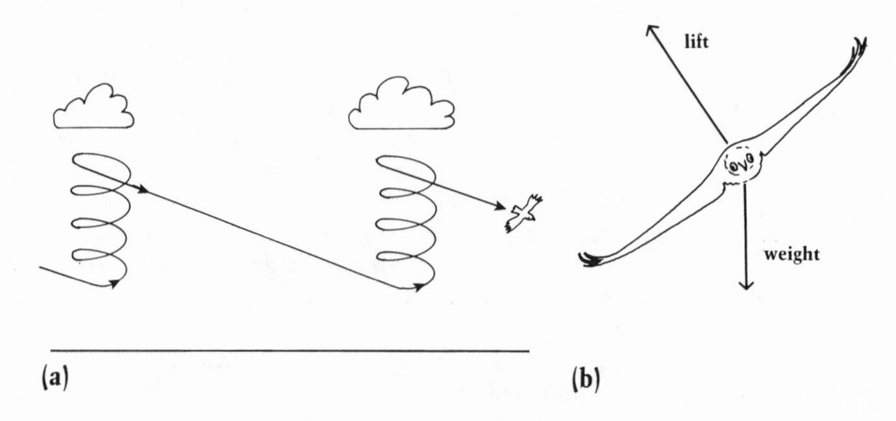

FIGURE 8.7. Diagrams of thermal soaring.

However, we already know that

$$\text{Maximum lift} = \text{constant} \times Av^2$$

where A is the wing area. These two statements tell us that

$$(\text{constant} \times Av^2) \text{ is greater than } mv^2/r.$$

Cancel out v^2 from both sides of this statement and rearrangement it to get r on the left hand side

$$r \text{ is greater than } [(m/A) \div \text{constant}].$$

This tells us that it is impossible to glide in circles of less than a certain radius. The minimum possible radius is proportional to m/A and therefore to wing loading. Thermal soarers, that have to glide in small circles, should have low wing loadings.

These arguments tell us that slope soarers should have high wing loadings and thermal soarers should have low ones. Earlier in the chapter I argued that large birds will have larger wing loadings than similar-shaped small birds. Figure 8.8a seems to show that all these things are true. It shows that slope soarers have larger wing loadings than thermal soarers of equal mass but that within each group large birds have larger wing loadings than small ones. *Pteranodon* has a remarkably low wing loading for its mass, lower even than for the modern thermal soarers, whether you use the wide-winged or the narrow-winged reconstruction.

Wing loading is only one of several features of aircraft that affect gliding performance. Another is aspect ratio, the ratio of wing span to mean chord. Figure 8.4 defines these terms and contrasts the high as-

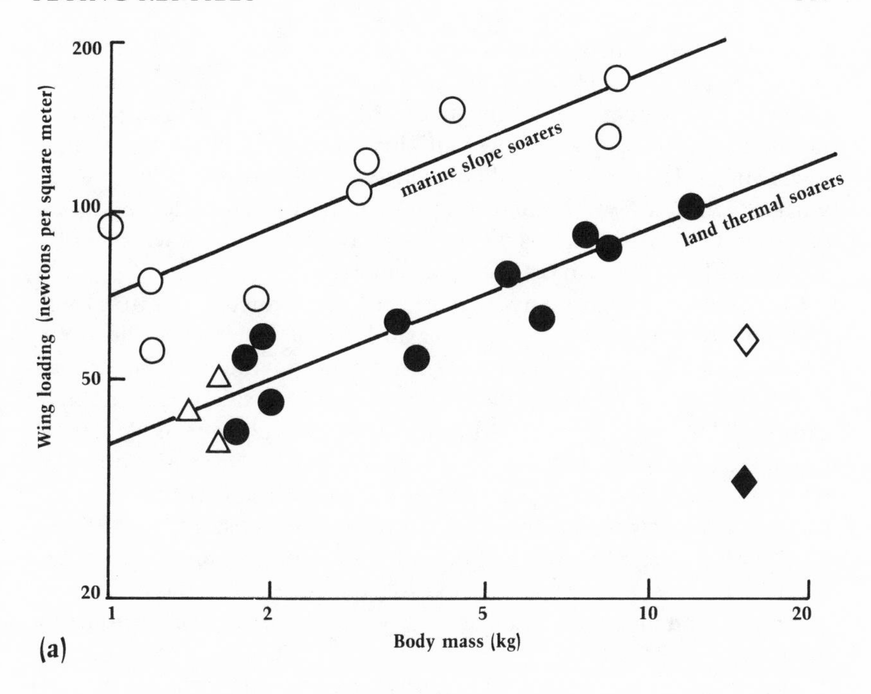

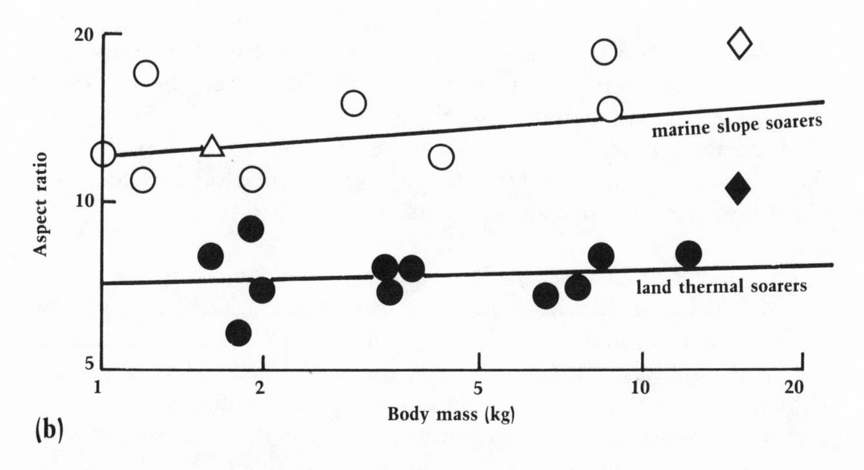

FIGURE 8.8. (a) A graph of wing loading against body mass and (b) a graph of aspect ratio against body mass for various soaring birds and *Pteranodon:* ○ marine slope soarers (albatrosses, shearwaters, petrels, gulls); ● land thermal soarers (storks, vultures, eagles); △ marine thermal soarers (frigate birds); ◆ *Pteranodon* (wide-winged reconstruction); ◇ *Pteranodon* (narrow-winged reconstruction).

Data from Brower 1983 and Padian 1985.

pect ratio (long, narrow) wings of an albatross with the low aspect ratio (relatively short, broad) wings of a vulture. As a general rule, gliders with high aspect ratios perform better than ones with lower aspect ratios: they are capable of lower sinking speeds and of gliding at shallower angles. However, it would obviously not be sensible to build wings with a span of a few kilometers and a chord of a few millimeters: excessively large aspect ratios would mean awkwardly long wings which would be difficult to make strong enough.

I am going to explain why high aspect ratios are best, but first I will have to explain one of the most basic laws of mechanics, Newton's Second Law of Motion. The momentum of a moving body is its mass multiplied by its velocity. (Please do not confuse this linear momentum with the angular momentum mentioned in chapter 5.) A force is needed to change the momentum of a body, and Newton's Second Law says that the force equals the rate of change of momentum. Acceleration is rate of change of velocity, so if mass is constant, mass times acceleration is rate of change of momentum. For most purposes we can express the law in its most familiar form, force equals mass times acceleration, but in this section it is more convenient to mention momentum.

The wings of an aircraft get lift by pushing on the air they pass through, giving it momentum (figure 8.3a). If they push on a mass M of air in each unit of time, giving it downward velocity w, the momentum given to the air in each unit of time is Mw. This is the rate of change of momentum and is equal to the lift force. The air is also given kinetic energy, amounting to $\frac{1}{2}Mw^2$ in each unit of time. This kinetic energy must be supplied somehow, either from work done by the engines of a powered airplane or from potential energy lost as a glider loses height. Aircraft designers want to keep it as small as possible.

You can get the necessary lift either by pushing on a little air, giving it a high velocity (low M, high w) or by pushing a lot of air to a low velocity (high M, low w). The latter requires less energy. You can get the same Mw for less $\frac{1}{2}Mw^2$ if you make M large and w small. An airplane will need less power and a glider will lose height less fast, if they are designed so that their wings push on as much air as possible.

Figure 8.3d shows the air that the wings push on. The bigger the wing span, the more air gets pushed. High aspect ratio (long, narrow) wings have larger spans than low aspect ratio ones, for the same wing area. That means that, in general, high aspect ratios are best.

Figure 8.8b shows aspect ratio plotted against body mass. The slope soaring albatrosses etc. have high aspect ratios and so also do the frigate birds, the marine thermal soarers. However the storks, vultures, etc., that soar in thermals over land, have relatively low aspect ratios. One possible reason is that longer, higher aspect ratio wings, that would give better gliding performance, might be awkward when the bird was taking off from land. The point for the wide-winged reconstruction of *Pteranodon,* in figure 8.8b, is between the lines for slope soarers and land thermal soarers, but the point for the narrow-winged one is above the line for slope soarers.

Whether the wings of *Pteranodon* were broad or narrow, the wing loading is low enough to suggest that it soared in thermals. It is not so clear whether it soared inland or over the sea. The wide-winged reconstruction is not too different from the large land soarers, with an aspect ratio only a little higher than theirs and with an even lower wing loading than any of them have. The narrow-winged model seems like a larger version of the frigate birds, marine soarers with high aspect ratios and low wing loadings. I have already said that I prefer the narrow-winged reconstruction, on anatomical grounds. Now we will test the hypothesis, that *Pteranodon* was a marine thermal soarer.

The hypothesis suggests that *Pteranodon* fossils should be found in rocks formed from sediments laid down in the sea. They are actually found in the central and western United States and southern Russia, mostly in places that are now well away from the sea. However, the same rocks contain fossils of sharks and other fishes, plesiosaurs (see Chapter 9) and turtles so it seems clear that they were formed in the sea and that coastlines have been moved around by subsequent earth movements.

This is not enough to show that the hypothesis is plausible. It is only in the trade wind zones that there are likely to be enough thermals over the sea for soaring to be feasible. Nowadays the trade winds blow only in and near the tropics, from about 25°N (the latitude of Miami) to 25°S (a little south of Rio de Janiero). *Pteranodon* is found further north. It is particularly abundant in Kansas (about 40°N) but is also found in Alberta (55°N). The continents have moved a little over the earth's surface since the Cretaceous period, but the latitudes of the *Pteranodon* sites have not been changed much. There is a lot of evidence that climates were generally warmer than they are now, but I have not been able to find any reliable opinion as to how far north it would have been possible to soar in thermals over the sea.

The hypothesis, that *Pteranodon* was a marine thermal soarer, also suggests that it would have eaten food from the sea. Fish scales and bones have been found in some fossils, in the position of the stomach. Another fossil had remains of fish and a leg of a crustacean in its throat. *Pteranodon* probably fed mainly on fish, and its bird-like beak seems suitable for catching them, though perhaps rather narrow.

Pteranodon seems much too fragile to have dived into the water to catch fish. It seems much more likely that it fed like frigate birds, flying low over the water and grabbing fish or squid without landing, putting only its beak into the water. The narrowness of the beak might actually be an advantage, because it would reduce the drag of the water on it. *Pteranodon* may have kept itself airborne by thermal soaring, searching for prey swimming close to the surface of the sea and swooping down to catch them.

The evidence does not prove that *Pteranodon* flew and fed like a giant frigate bird, but it seems consistent with the idea. Thermals over the sea continue day and night, in the trade wind zone, so *Pteranodon* could have soared for days on end without landing, seldom having to flap its wings. It could not have flown very fast because of its low wing loading, but if a storm blew up it could have avoided being blown downwind by landing on the surface of the sea.

The long beak may have been good for catching fish but seems likely to have been troublesome in flight. All would be well so long as it faced directly forward, but if it was turned to either side the air would press on one side of the beak and try to twist the head round like a weather cock. *Pteranodon* would need strong neck muscles, unless it had some way of canceling the effect out.

Cherrie Bramwell and G. R. Whitfield of the University of Reading thought that the crest on the back of *Pteranodon*'s head (figure 8.2b) might have helped to cancel out the twisting effect, keeping the head facing forward. They tested the idea by making models of the head, with and without the crest, and putting them in a wind tunnel. They set them at various angles to the wind and measured the moments that the aerodynamic forces exerted on them. They found to their surprize that the crest had little effect.

Different species of *Pteranodon* have the crest set at different angles and most pterosaurs (including some large ones with long beaks) have no crest. It seems possible that the main function of the crest had nothing to do with aerodynamics. It may have been an ornament with the same sort of function as the crests of hadrosaurs (chapter 5). It has also been suggested that it may have been a cooling fin like the plates on the backs of stegosaurs (chapter 7).

This chapter has concentrated on *Pteranodon,* the largest flying animal known from reasonably complete fossils. It was enormous, with a wing span of 7 meters, but very lightly built. We know the span of its wings, but paleontologists disagree about their shape (figure 8.2a). Whether the wings were wide or narrow, its wing loading was apparently low, which would have enabled it to fly slowly and suggests that it soared in thermals. The wide wings of one restoration resemble the wings of vultures, which soar in thermals over land. The narrow wings of the other are like those of frigate birds, which soar in thermals over the sea in the trade wind zone. It seems likely that *Pteranodon* flew like frigate birds and fed like them, grabbing fish from the surface of the sea.

Principal Sources

Bramwell and Whitfield (1974) investigated the aerodynamics of the wide-winged restoration of *Pteranodon* and Brower (1983) did the same for the narrow-winged restoration. Padian (1983, 1985) argued that the narrow-winged restoration was the more realistic. Pennycuick (1983) investigated the flight of frigate birds and Rayner (1987) has shown how the wing loadings and aspect ratios of birds are related to their styles of flying. Wellnhofer (1975) is the source for figure 8.1.

Bramwell, C. D. and G. R. Whitfield. 1974. Biomechanics of *Pteranodon. Philosophical Transactions of the Royal Society* B 267: 503–581.

Brower, J. C. 1983. The aerodynamics of *Pteranodon* and *Nyctosaurus*, two large pterosaurs from the Upper Cretaceous of Kansas. *Journal of Vertebrate Palaeontology* 3:84–124.

Currey, J. D. and R. Mc N. Alexander. 1985. The thickness of the walls of tubular bones. *Journal of Zoology* A 206:453–468.

Padian, K. 1983. A functional analysis of flying and walking in pterosaurs. *Palaeobiology* 9:218–239.

Padian, K. 1985. The origins and aerodynamics of flight in extinct vertebrates. *Palaeontology* 28:413–433.

Pennycuick, C. J. 1983. Thermal soaring compared in three dissimilar tropical bird species, *Fregata magnificens, Pelecanus occidentalis* and *Coragyps atratus. Journal of Experimental Biology* 102:307–325.

Rayner, J. M. V. 1987. Form and function in avian flight. *Current Ornithology* 5: 1–66.

Wellnhofer, P. 1975. Die Rhamphorhynchoidea der Oberjura Plattenkalke Suddeutschlands. *Palaeontographica* A 148:1–33, 132–186, and 149:1–30.

IX

Marine Reptiles

WE WILL now discuss the giant reptiles that lived in the sea in the Mesozoic era, when the dinosaurs were living on land. They had flippers instead of feet and would probably have been pretty helpless on land. We know that they lived in the sea rather than in fresh water because the other fossils found in the same rocks include sea urchins and squid-like molluscs, members of groups whose modern members are found only in the sea.

I am going to start with the ichthyosaurs—reptiles that looked remarkably like fish (figure 9.1). Most of them were at least a meter long and some were as much as 15 meters. I have a model of one of the larger kinds, *Ichthyosaurus,* bought from the Natural History Museum, London. I measured its volume and calculated the mass of the living animal in the same way as for dinosaurs (chapter 2). This particular animal was 8 meters long and I calculate that its mass was 6 tonnes. Adult Killer whales have about the same length and mass.

Most ichthyosaur fossils are skeletons and nothing more, but a few have dark marks showing the outline of the body and of the fins and flippers. These marks show that the flippers were a good deal broader than you might guess from the skeletons, and that at least some ichthyosaurs had a fin on the back. Some ichthyosaurs had straight tapering tails but the best known kinds, including the one in figure 9.1, had a sharp kink in the backbone where it entered the tail fin. Dark marks on a few fossils show that these ichthyosaurs had tails shaped like crescent moons.

There are two groups of modern fish that look very like these ichthyosaurs: the tunnies and the porbeagle sharks (figure 9.2). As well as being shaped like ichthyosaurs they overlap the ichthyosaur size range. For example, Bluefin tuna reach lengths of 4 meters and masses of 0.8 tonnes. Great white shark grow to maximum lengths of about 11 me-

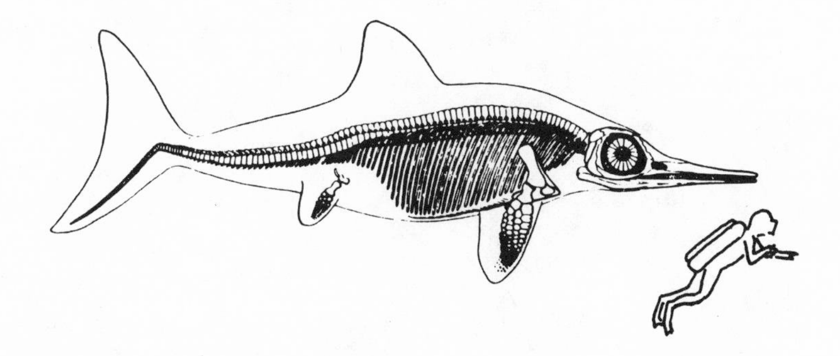

FIGURE 9.1. The skeleton of a Jurassic ichthyosaur, with an outline of the body. From Romer 1966. I have added a frogman to scale, assuming this ichthyosaur is 8 meters long.

ters. Whales also look like ichthyosaurs, but with a very obvious difference. Ichthyosaurs had vertical tail fins (like tunnies and sharks and presumably beat them from side to side when they swam. Whales have horizontal tail flukes and beat them up and down.

In one respect ichthyosaurs were even more like dolphins than like tunnies or sharks: they had long narrow jaws with a lot of simple pointed teeth. Dolphins eat fishes and squid, and ichthyosaurs seem to have eaten similar things. Many ichthyosaurs have been found with their fossilized stomach contents still in place inside them, enclosed by their ribs. Some fish scales have been found in them, and enormous numbers of hooks from the suckers of squid-like mollusks.

Tunnies, porbeagle sharks, and whales, the modern animals shaped like ichthyosaurs, swim fast. The most reliable speed measurements have been made with dolphins trained to swim as fast as possible over a marked course, or to follow a lure towed by a fast boat. The highest speed on record seems to be 11 meters per second (25 mph) for a Spotted porpoise 2 meters long, and slightly slower speeds have been recorded for other species. These are sprint speeds, maintained for only a few seconds. They are astonishingly fast for movement in water, and equal the top speeds (on land) of human sprinters.

Speeds of tunnies have been measured, both by filming them and by catching them on a rod with an instrumented reel, that recorded the rate at which the fish pulled out the line. Several records of 5 to 13 meters per second were obtained in this way, and also two of 21 meters per second for a Wahoo and a Yellowfin tuna. I find these last two records hard to believe because they are so much higher than any others. I wonder whether some error was made: for example, a mistake could conceivably have been made about the speed at which the re-

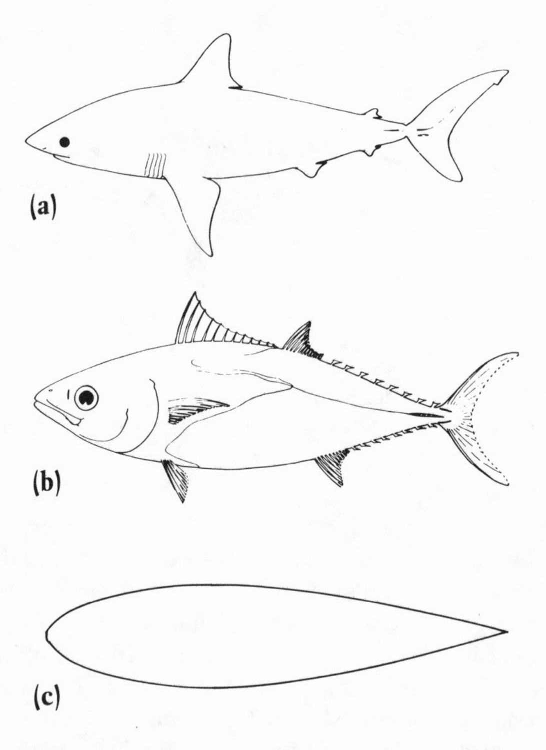

FIGURE 9.2. A Porbeagle shark, a Bluefin tuna, and an optimum streamlined shape. The fishes are from drawings by Valerie du Heaume in Wheeler 1969.

cording equipment was running. Even if we reject these records (as I am inclined to do) it seems clear that tunas, like dolphins, swim very fast. The fastest shark speed I have seen recorded is 5 meters per second, but there are very few data.

The speeds of the similar-shaped modern animals make it seem likely that ichthyosaurs were also fast. My guess is that they could have sprinted at 10 meters per second.

Drag resists the movement of bodies through water, as also through air. It is much larger in water than at the same speed in air because water is so much denser. To minimize drag, a body should be designed to disturb the water as little as possible, as it passes through. Any body will leave a wake of swirling water behind it, but the narrower the wake, the less the drag. This is because energy is needed to set the water swirling: the kinetic energy of the swirling water comes from work done against drag. Streamlining is the art of designing bodies so as to disturb the water as little as possible. The best shapes are rounded in front, and taper to a fine point behind to allow the water to close in smoothly after the body has passed. Torpedoes and submarines are shaped like this, and so also are ichthyosaurs, whales, and many fish.

When airships were used, the engineers who designed them set out to discover the best shape. The carrying capacity of an airship depends on its volume because it is supported by the buoyancy of gases that are lighter than air, so the basic problem was to find the shape that gave least drag for given volume, at any particular speed. The answer turned out to be a streamlined shape with the length 4.5 times the diameter at the fattest part (figure 9.2, bottom).

The same shape seems likely to be best for swimming animals, so it is not surprizing to find that ichthyosaurs, tunnies etc. are very nearly this shape. The *Ichthyosaurus* model (already mentioned) has its length 5.0 times the maximum diameter, Yellowfin tuna are 4.5 diameters long, and Porbeagle sharks and Bottle-nosed dolphins are both about 5.5 diameters long. (The diameter that I have used in these calculations is the mean of the maximum height of the body and the maximum width.)

Figure 9.3 shows how tunnies swim. They beat their tails from side to side as they move forward, so the tail takes a wavy path through the water. It is held at an angle of attack so that lift acts on it as well as drag. (Lift acts on hydrofoils in water, just as on aerofoils in air.) While the tail is moving to the right, the lift acts forward and to the left. While it is moving to the left, the lift acts forward and to the right. The components to left and right cancel out, over a complete cycle of tail movements, so the net effect is a forward thrust, driving the fish through the water. The drag on the tail acts backward all the time, reducing the thrust, but if the hydrofoil is well designed (as tunny tails seem to be) the drag is relatively small. Ichthyosaurs like the one in figure 9.1 presumably swam like this but it has been suggested that some of the ones with narrow tapering tail fins may have depended more on flipper movements.

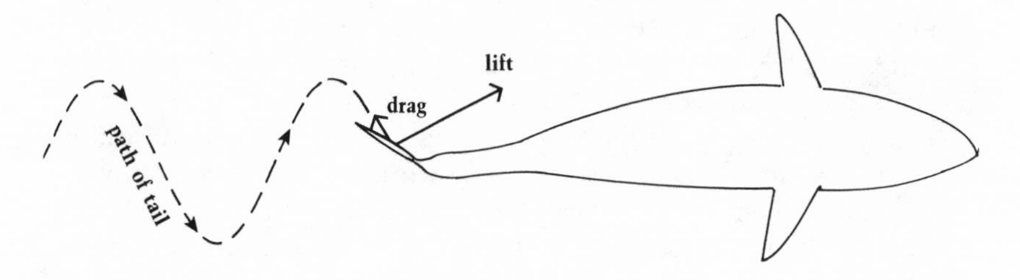

FIGURE 9.3. How tunnies swim. Ichthyosaurs presumably swam in the same way.

The aspect ratios of tails can be measured in the same way as for wings, by dividing the span of the tail by its mean chord (figure 8.4). High aspect ratio tails give the same lift for less drag, like high aspect ratio wings, and make for efficient swimming. Tunny tails have high aspect ratios, for example 7 for Yellowfin. Whale flukes have lower aspect ratios, for example 5 for White-sided dolphin, possibly because they have no skeleton to stiffen them and might be too flexible if their spans were increased. Tail outlines preserved in ichthyosaur fossil show still lower aspect ratios, for example 3.7 for *Ichthyosaurus.* In this respect ichthyosaurs seem markedly inferior to tunnies.

As well as being similar in shape, tunnies, porbeagle sharks, and whales have a remarkable thing in common. All of them are endotherms. You would expect whales to be endotherms, because they are mammals, but tunnies and porbeagle sharks seem to be the only endothermic fishes. They do not keep themselves as warm as mammals, and they do not heat the whole body to the same temperature: the warmest parts deep inside the body are seldom above 35°C and often much cooler, whereas mammals keep their bodies at 36–40°C. However, these warmest parts are often 10 to 15 degrees warmer than the water in Bluefin tuna, and 10 degrees in Porbeagle sharks. It seems quite likely that ichthyosaurs were also endotherms.

Dolphins often leap repeatedly from the water, as they swim: this is called "porpoising." It looks as if they are playing or exercising, but it has been suggested that they may actually save energy by this apparently strenuous behavior. The main point is that drag is *very* much less in air than in water, so if the dolphins can make part of their journey out of the water they may save energy.

I will try to explain the idea in a bit more detail, because it seems possible that ichthyosaurs also porpoised. Obviously, it costs energy to leap: the amount of energy is proportional to the height of the leap. The length of the leap depends on the dolphin's speed and on the angle at which it leaves the water, but faster take-off (at any particular angle) makes the leap longer and also higher. If leaps at the same angle are compared, the height of the leap (and so the energy needed) is proportional to the length of the leap, so the energy needed per meter leaped is the same for all swimming speeds.

A leap saves energy if the work needed for it is less than would be needed to swim the same distance. The work needed for swimming is done against drag, and can be calculated by multiplying the drag by the distance. (The work done against a force is the force multiplied by the distance moved against it.) Drag is roughly proportional to the square of speed, so the work needed to swim a meter increases with increasing speed, while the work needed to leap a meter remains constant. As a dolphin swims faster and faster it must eventually reach a speed at which leaping saves energy.

People have tried to calculate the critical speed, and one estimate for dolphins is 5 meters per second, which is well within their range of speeds. However, there are all sorts of doubts about the calculation, both about the amount of work needed for leaping and about the amount needed for swimming. All we can feel sure of is that there must be some speed, above which leaping would save energy if dolphins can swim that fast. If dolphins leap for this reason, perhaps ichthyosaurs leapt too. I like to imagine that they did, but tunnies give no support to the idea. They do not leap although they are much the same size and shape as dolphins and seem able to swim as fast.

Most bony fish have a gas-filled float (the swimbladder) in the body cavity, making their densities about the same as the density of the water they swim in. Sharks and most tunnies have no swimbladder, so they are denser than water and would sink if they stopped swimming. Porbeagle sharks and tunnies live near the surface of the oceans, often with the bottom thousands of meters below, so there is no question of stopping for a quick rest on the bottom. These fish swim all the time, and some apparently have to keep swimming in any case, to get the oxygen they need. (Instead of making breathing movements like other fish they simply swim around with their mouths open, letting the water flow through their gills.) They prevent themselves from sinking mainly by keeping their pectoral fins spread like airplane wings, at a suitable angle of attack to give the necessary lift. (The pectoral fins are the large pair close behind the head.)

Whales are less dense, partly because they have a lot of blubber (fat is slightly less dense than water) but largely because they have air-filled lungs. Whales can often be seen floating at the surface of the sea with their backs protruding slightly above the surface, showing that they are actually less dense than the water. Their buoyancy enables them to stop swimming and rest without sinking. Ichthyosaurs presumably also had lungs (it would be surprising to find a reptile without them) and probably also had densities close to the density of water. Similarly crocodiles have about the same density as (fresh) water, as I showed in chapter 2. Ichthyosaurs probably had no need to keep swimming like tunnies and Porbeagle sharks, but could stop and rest like whales.

Ichthyosaurs must have come to the surface regularly to breathe, but they may also have dived to substantial depths, like whales. If they had the same density as water while at the surface, they would be denser while diving, because the air in their lungs would be compressed to a smaller volume. The pressure at a depth of 10 meters is twice the pressure at the surface and would halve the volume of the air; at 20 meters the pressure is three times as much as at the surface and the volume of the air would be reduced to one third; and so on. During a deep dive the lungs would give very little buoyancy and the animal would sink if it stopped swimming. It might get the necessary lift, as it swam, by spreading its flippers.

Ichthyosaurs may have had to dive quite deep, to get their food. Their skulls show that they had large eyes, which suggests that they depended on sight to find prey. Further, it suggests to me that they fed by day (but I admit that big eyes could be an adaptation for feeding in dim light, at dusk). Many fish and squids spend the night near the surface but dive quite deep by day: for example, herring that spend the night near the surface often dive to 100 meters or more. If ichthyosaurs fed by day on prey that behaved like that, they would have had to dive.

The ichthyosaurs seem splendidly adapted for swimming, but they would probably have been as helpless on land as a stranded whale. They could hardly have crawled up beaches to lay eggs above the high tide mark, as the sea turtles do. They could not have laid their eggs in water, because the embryos would have suffocated, as the embryos of birds and modern reptiles do if their eggs are submerged. The reason is that oxygen diffuses much more slowly through water than through air. If the pores of the eggshell get filled with water, oxygen cannot diffuse in fast enough.

Ichthyosaurs seem to have got round this problem by giving birth instead of laying eggs. Adults have been found with up to 12 young ones fossilized inside them, enclosed by their ribs. It is possible that they had eaten the little ones, but it is pleasanter to think that they were pregnant. Though most modern reptiles lay eggs, some sea snakes and others give birth.

The mosasaurs were another group of marine reptiles, but they looked much less fish-like than the ichthyosaurs, more like crocodiles with flippers instead of legs. The details of their skulls show that they were actually lizards, closely related to the Komodo dragon and other monitor lizards. The biggest of them were at least 9 meters long, about the same as the biggest modern crocodiles. (Though the Komodo dragon is the largest modern lizard it grows little more than 3 meters long.) The mosasaurs lived in the Cretaceous period, at the end of the Mesozoic era.

I speculated that ichthyosaurs may have dived, and there is some evidence that mosasaurs did. X-ray pictures of many of their vertebrae show the kind of damage that would have occurred if the blood supply to parts of the bone had got cut off, while the animal was still alive.

It has been suggested that the damage may have been caused by the bends, a serious hazard of diving.

Here is how the bends happens to human divers. The air they breathe has to be compressed to match the pressure of the water where they are working. The extra pressure makes extra gas dissolve in the blood. When the diver returns to the surface the extra gas comes out of solution, forming bubbles that may block blood vessels, causing damage, pain and even death. Human divers avoid the bends by returning slowly to the surface but a diving mosasaur would have to get to the surface reasonably soon, for its next breath. Whales avoid the bends largely by having small lungs and a big windpipe: the high pressures that they meet when they dive collapse their lungs, forcing the air back into the windpipe whether there is less danger of too much gas being absorbed into the blood.

The plesiosaurs lived in the Mesozoic seas, like ichthyosaurs and mosasaurs. Also like ichthyosaurs and mosasaurs they had two pairs of flippers instead of the fore and hind legs of most other reptiles. The shapes of their bodies, however, were quite different from those of the other marine reptiles. The trunk was broad and relatively low, a less extreme version of the body shape seen in modern turtles. Attached to it was either a long neck with a small head, or a short neck with a relatively large head (figures 9.4, 9.5). Some of them were very large, as the picture shows, but even *Elasmosaurus* was only half as long as

FIGURE 9.4. A short-necked plesiosaur (*Kronosaurus*) and a long-necked one (*Elasmosaurus*) from Romer 1968, with a frogman.

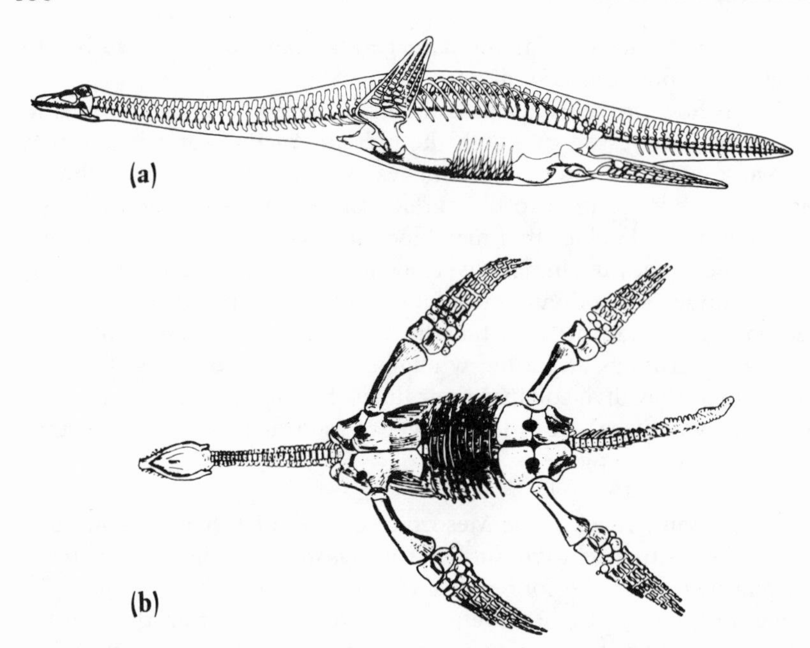

FIGURE 9.5. Skeletons of two long-necked plesiosaurs. *Cryptoclidus* (length 4 meters) is shown in side view and *Thaumatosaurus* (length 3.4 meters) is shown from below. From Brown 1981, by courtesy of the British Museum (Natural History), and Romer 1966, respectively.

the dinosaur *Diplodocus.* The Natural History Museum, London, sells models of a very large (14 meter) long-necked plesiosaur. One of my students, Debbie O'Hare, measured its volume and calculated that the living animal had a mass of 7.5 tonnes (or perhaps a little less: the model is possibly a little too deep in the body). In comparison with that, the Leatherback, the biggest of the modern sea turtles, grows to shell lengths of only 2 meters and masses of about 0.6 tonnes.

Plesiosaurs seem to have had small tails that would not be very effective for swimming. They presumably propelled themselves mainly by flipper movements, either by rowing (like freshwater turtles) or by "underwater flight" (like marine ones).

Figures 9.6a and 9.7a show how plesiosaurs may have rowed. They show the flippers moving backward and forward, like the oars of a boat. In the backward stroke the blades of the flippers are held vertical, so as to push as hard as possible against the water. In the forward stroke they are held horizontal so as to strike the water edge-on and meet as little resistance as possible. (A human rower would lift the oars out of the water for the forward stroke, but an animal swimming below the surface cannot do that.) Notice that in the power stroke the flipper blade is moving backward through the water so the drag on it (the force opposite to the direction of movement) acts *forward.* Rowing boats and rowing animals are propelled by forward-acting drag on backward-moving oars.

Figures 9.6b and 9.7b show a different method of swimming, which is called underwater flight because the movements are like those of flying birds. Penguins, which cannot fly in air, use their wings in this way to swim underwater. Sea turtles also swim this way, using their flippers. The flippers beat up and down. On the downstroke they are held at an angle of attack so that lift acts on them, forward and upward (figure 9.7b). For the upstroke the angle is adjusted so that the lift acts forward and downward. The upward and downward components cancel out over the complete cycle of movements so that the net effect is a forward thrust (which is reduced a bit by the drag on the flipper). Notice how similar figure 9.7b is to the diagram of a tunny swimming in figure 9.3. Tunnies and presumably ichthyosaurs swim by means of their tails, beating them from side to side, and turtles and possibly plesiosaurs swim by means of their flippers, beating them up and down, but the basic principle is the same in both cases.

I would like to emphasize that rowing and underwater flight are utterly different techniques. In rowing, the flippers or oars are moved backward and forward and the propulsive thrust comes from the drag on them in their backward strokes. In underwater flight the movement is up and down and the thrust comes from lift: in this case, the drag is simply a hindrance.

I have been assuming that plesiosaurs had about the same density as water, so that an upward force at one stage of the cycle of flipper movements must be balanced by a downward force at another. In figure 9.7b this balance comes from an upward-sloping force in the downstroke and a downward-sloping force in the upstroke. Figure 9.7c shows another possible way of avoiding unbalanced vertical forces. The flipper is moved almost vertically downward, held at an angle of attack so as to give forward lift. It is then raised on a sloping path, moving edge-on, with no angle of attack, so that the forces on it are very small. The downstroke gives horizontal thrust and the upstroke very little force. Figure 9.6c shows on the left how the flippers would have to move *relative to the animal's trunk,* to follow the appropriate path through the water: they would have to beat down and back, then forward and up. The thrust, in this swimming technique, comes from lift, so it is a form of underwater flight. Sea lions seem to swim rather like this.

Plesiosaurs would have had to drive water backward, to propel them-

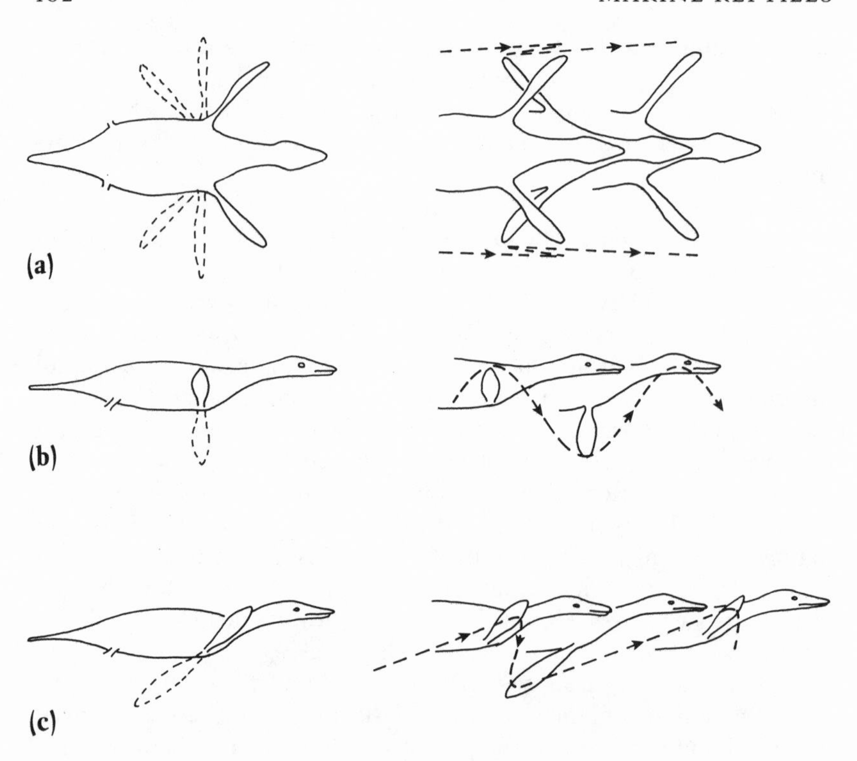

FIGURE 9.6. Three possible swimming techniques for plesiosaurs: (a) rowing; and (b) and (c) underwater flight. The diagrams on the left show how the flippers would have been moved relative to the body and those on the right show successive positions of the animal moving through the water. Only the fore flippers are shown.

selves forward. In rowing, the flippers would push fairly small lumps of water backward (figure 9.8a). In underwater flight the flippers, beating up and down through a large angle, would affect much more water (figure 9.8b). When I wrote about the aspect ratios of wings (chapter 8) I explained that less energy is needed to get a force by accelerating a large mass of fluid to a low velocity, than by accelerating a small mass to a high one. This argument says that underwater flight should be more economical than rowing. It should need less energy for swimming at the same speed.

That is one reason for thinking underwater flight more likely than rowing. Animals tend to evolve efficient ways of doing things. There are animals that row, but at least some of them (freshwater turtles, and ducks) use their legs for walking as well as for rowing. It would

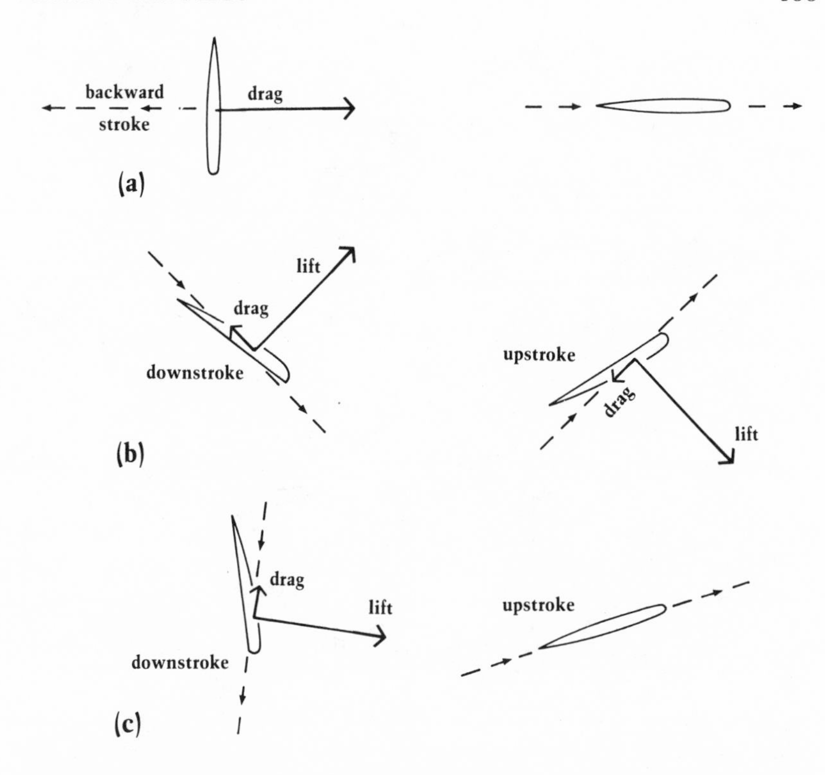

FIGURE 9.7. Diagrams showing a section through a flipper, and the forces acting on it, at different stages of the three swimming techniques of figure 9.6.

be very difficult to design a foot which was both effective for walking on land and suitably streamlined for use in underwater flight.

Another reason for thinking underwater flight more likely is that plesiosaur flippers taper at the tips. There is no advantage in tapering the tip of an oar blade, and the oars used in rowing races are made with square ends. However, there is an advantage in giving aerofoils and hydrofoils tapered, rounded ends: such shaping can spread the lift out over the span of the hydrofoil in the best possible way, so as to get lift with as little drag as possible. Bird and airplane wings, and propeller blades, generally taper toward their tips. The shape of plesiosaur flippers suggests that their function was to provide lift, not drag.

For underwater flight, plesiosaurs would have to have been able to flap their flippers up and down. The shapes of the joints seem to show that they could have done this, though they probably could not have raised the flippers very high above their backs. They would also have

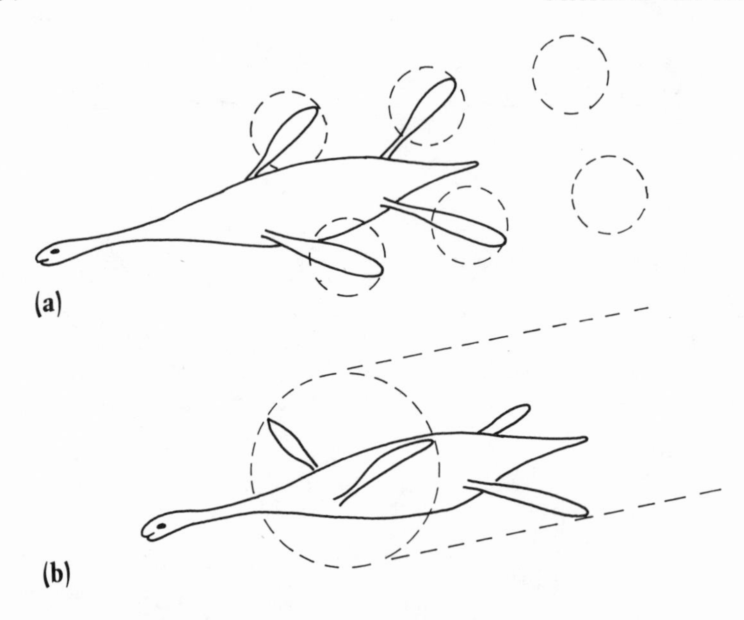

FIGURE 9.8. A plesiosaur (a) rowing and (b) "flying" under water. Broken outlines show the water driven backward by the swimming movements.

needed appropriate muscles. Scientists have tried to work out how the flipper muscles were arranged, by looking at plesiosaur skeletons. Figure 9.5 shows big plates of bone to which muscles could have attached in the chest (between the fore flippers) and on the underside of the abdomen (between the hind flippers). These are the ventral parts of the pectoral and pelvic girdles. They seem excellent areas of attachment for muscles that would pull the flippers down, in a powerful downstroke, but there does not seem to be much to attach upstroke muscles to. The big plates of bone are all below the shoulder and hip joints. The upward extensions of the pectoral and pelvic girdles (seen in the side view) seem to be attached rather weakly to the ribs and backbone. Strong upstroke muscles could have been attached to the backbone, but there is little to prevent them pulling the girdles bodily upward instead of flapping the flippers. The symmetrical style of underwater flight, shown in figures 9.6b and 9.7b, needs equally strong upstroke and downstroke muscles. The style shown in figure 9.6c and 9.7c needs big forces for the downstroke but only small forces for the upstroke, and seems the more likely swimming technique for plesiosaurs.

Plesiosaurs probably could not swim very fast. One reason for thinking this is that the volume of flipper muscle that it seems possible to fit into their bodies is relatively small, compared to the volume of tail muscle that would be found in similar-sized fishes or whales. The swimming muscles of a 43-kilogram porpoise had a mass of 9 kilograms. (This includes the tail muscle and most, but not all, of the back muscle.) That is 21 percent of body mass. I do not see how the flipper muscles of plesiosaurs could have been as much as 21 percent of body mass.

Another reason for thinking that plesiosaurs were probably rather slow is that the swimming technique we suspect they used (figure 9.7c) requires the flipper to be moved almost vertically down through the water. To do this while the animal was moving forward, the flippers would have to move backward (relative to the body) as fast as the body was moving forward (relative to the water). Imagine the plesiosaur shown in figure 9.7c swimming at 10 meters per second, about the maximum sprinting speed of tunnies and dolphins. During the power stroke, a point at the center of the flipper would have to move backward through a distance x at about 10 meters per second. If the animal was a moderate-sized short-necked plesiosaur, three meters long, x would be a meter or less and the movement would have to be made in one-tenth of a second. If the upstroke was made at the same speed, the cycle of flipper movements would be completed in 0.2 seconds and the flapping frequency would be $1/0.2 = 5$ cycles per second. I doubt whether so large an animal could have managed so high a frequency of movement. Small penguins beat their wings when they swim at up to 4 cycles per second, but they are very much smaller. There is a general rule that large animals cannot move their limbs at as high frequencies as small ones: a horse cannot make as many strides per second as a mouse, and a swan cannot make as many wingbeats per second as a sparrow. Figure 9.9 shows some data. Penguin wing beat frequencies are about the same as the wing beat frequencies of similar-sized flying birds (although penguins move their wings in water) but considerably less than the stride frequencies of similar-sized mammals, which is not surprizing: the penguins were moving their wings in water. The points are widely scattered around the lines but the general trend is clear: bigger animals use lower frequencies. By extending the penguin line, I estimate that an underwater flier 3 meters long (the size of the plesiosaur I have been discussing) would have beaten its flippers at a maximum frequency of about 1 cycle per second. This is only one fifth as much as I estimated would be needed for swimming at 10 meters per second and suggests that the plesiosaur could only manage about 2 meters per second.

It is not just chance that makes big animals move their limbs at lower frequencies than small ones: there is a good mechanical reason. Imagine two animals of the same shape, one twice as long as the other (and twice as wide and twice as high). It is eight times as heavy as the

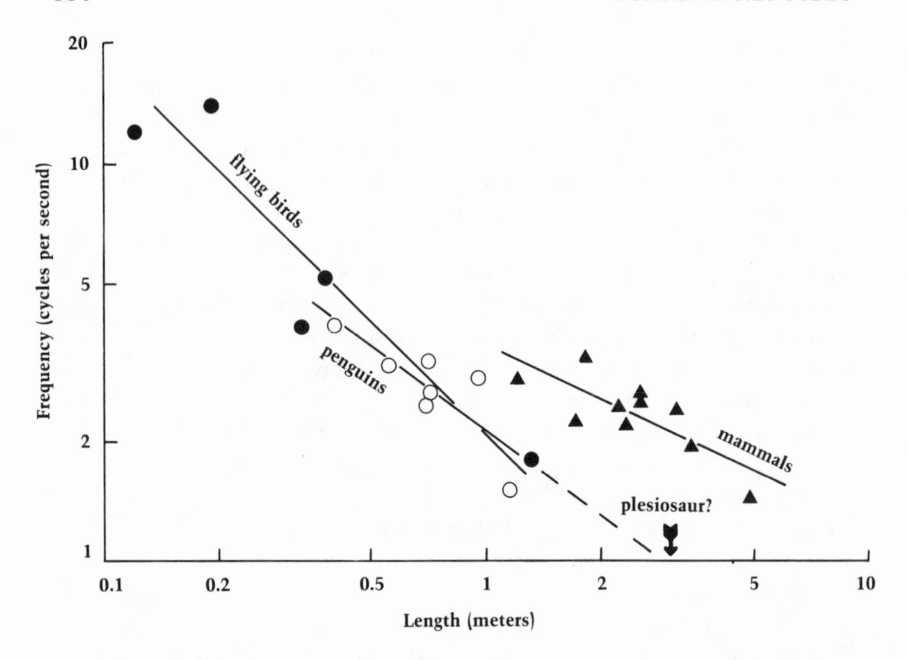

FIGURE 9.9. Bigger animals move their limbs at lower frequencies. The graph shows ●, mean wing beat frequencies of flying birds; ○, maximum wing beat frequencies of swimming penguins and ▲, mean galloping stride frequencies of hoofed mammals, all plotted against the overall length of the body. The penguin data come from Clark and Bemis 1979, the mammal data from my own observations in Kenya, and the flying bird data from various sources.

small one and has eight times as much limb muscle, able to do eight times as much work to accelerate the limbs at the beginning of each stroke. This work is used to give the limbs kinetic energy (half mass times speed squared) but its limbs, plus any water moved by them, have eight times the mass of the smaller animal's ones, so can only be accelerated to the same speed. The big animal's limbs have to travel twice as far as the small ones to make each stroke, so their cycle of movements takes twice as long. This argument says that doubling the length of an animal should halve its frequency of limb movements, and figure 9.9 shows that this is roughly true for flying birds and swimming penguins.

I have argued that the style of underwater flight shown in figure 9.6c is unlikely to be very fast, because the flippers have to move back, relative to the body, as fast as the body advances through the water. Penguins and sea turtles use more symmetrical styles, like the one shown in figure 9.6b which is not limited in this way, but even so they are not very fast. The highest speed shown in films of penguins swimming in Detroit Zoo was 3.4 meters per second (for a King penguin, about 90 centimeters long) and adult Green turtles seem to swim no faster than 2.0 meters per second.

Penguins and sea turtles move their left and right wings or fore flippers in unison. If they did not they would waste energy by swimming a slightly zigzag route. Plesiosaurs probably also moved the left and right flippers of a pair together. Sea turtles swim mainly with their big fore flippers, beating their small hind flippers only occasionally. Plesiosaurs had big hind flippers as well as big big fore ones and probably used both pairs about equally.

I have argued that the downstroke was the power stroke. If so, there might be an advantage in beating the fore and hind flippers out of phase with each other as shown in figure 9.8b. Fore-power strokes would alternate with hind-power strokes, keeping the animal moving at a steady speed. This might save energy; drag is about proportional to speed squared, so the average drag is greater when swimming at a fluctuating speed than when swimming steadily at the same average speed. However, there might be a very serious disadvantage in beating the fore and hind flippers alternately. The hind flippers might find themselves moving in water that had already been accelerated by the fore flippers. They might accelerate this water further, but it would be more efficient for them to work on different water. This is another application of the principle we have already met several times: it is more efficient to get thrust by accelerating a lot of water to a low velocity, than less water to a higher velocity.

The long-necked plesiosaurs had extraordinarily long necks, some of them longer than the whole of the rest of the body. If they swam under water with the long neck stretched out in front it would have been quite tricky for them to steer a straight course: if the animal accidentally veered slightly to one side, the water, striking the neck obliquely, would tend to make it veer more. This is the opposite to the effect of flights on an arrow, which tend to correct any deviation, pulling the arrow back to a straight course. The difference is that the neck has a big surface area in front of the animal's center of gravity and the flights have a big area behind the arrow's one. It seems possible that plesiosaurs often avoided this problem by swimming at the surface with their necks out of the water. More energy is needed to swim at the surface than to swim well submerged, because an animal at the surface pushes a bow wave in front of it like a boat, but this might not be a big disadvantage if the plesiosaur swam slowly, and in any case it would have to visit the surface frequently to breathe.

If plesiosaurs had eaten worms or clams, we might suppose that they used their necks to reach down to the bottom, dabbling like ducks or swans, but their spiky teeth seem more suitable for catching fishes and squid-like animals which would probably have been too active to be caught easily that way. It seems likely that they darted at prey, extending their long necks to catch things as they swam by. The movement could have been fast, if the neck was held out of water. Herons use their long necks to dart at fish, though they stand in the water instead of floating as plesiosaurs presumably did.

The fossil record seems to show that the plesiosaurs became extinct at the same time as the dinosaurs, 65 million years ago, but some people believe that rather similar animals are still living in Loch Ness, Scotland (figure 9.10). The picture is not very like a plesiosaur (notice the humped shoulders, and the central ribs in the flippers) but there is some resemblance.

There have been reports of the Loch Ness monster since the Middle Ages, and tremendous efforts have been made in modern times to get good evidence of its existence. The surface of the lake has been kept under observation, sonar (echo sounding) has been used, and thousands of underwater flash photographs have been taken at random in the hope that the monster will swim into view. Some surface photographs have

FIGURE 9.10. An impression of the Loch Ness monster, based on eyewitness accounts and (unclear) photographs. From Scott and Rines 1975. Reprinted by permission.

been taken that show monster-shaped images, but it never seems certain that the image is not a floating log or an odd pattern of ripples. Some sonar traces have detected unexplained objects about 15 meters long and there are a few hazy underwater photographs (hazy even after computer enhancement) that show shapes like the neck and flippers in figure 9.10. If the monster is 15 meters long and has the shape shown in the picture, it must weigh well over 10 tonnes. If there is one monster there must be several, or at least there must have been several until quite recently: no animal is immortal, and very small populations are in danger of dying out. If there really are monsters there, it seems odd that we have not got better evidence of their existence.

This chapter has been about three groups of fossil reptiles. The ichthyosaurs were beautifully streamlined, like tunnies and dolphins, and probably swam fast. They may have been endotherms (like tunnies), they may have dived deep (like dolphins), and they may have porpoised when swimming at the surface.

The mosasaurs were giant lizards with flippers instead of legs. Some have damaged vertebrae that look like symptoms of the bends, a hazard of diving.

The plesiosaurs used flippers to swim, probably by underwater flying rather than rowing. They probably swam rather differently from turtles and penguins, getting thrust only from the downstroke. If so, they must have been rather slow. Some had remarkably long necks, which may have been held out of the water and used for darting at prey.

As for the Loch Ness monster, I am not convinced that it exists. Are you?

Principal Sources

Much of my information on fish swimming and on endothermic fishes comes from Hoar and Randall (1978). The hypothesis about porpoising comes from Au and Weihs (1980) and the top speed for dolphins from Lang and Prior (1966). Clark and Bemis (1979) and Davenport and others (1984) describe how penguins and turtles swim. Robinson (1975) and Godfrey (1984) discuss plesiosaur swimming. Rothschild and Martin (1987) found the evidence of mosasaurs getting the bends. Scott and Rines (1975) and Witchell (1975) present the evidence for the Loch Ness monster. The other items in the list are sources for illustrations.

Au, D. and D. Weihs. 1980. At high speeds dolphins save energy by leaping. *Nature* 294:548–550.

Brown, D. S. 1981. The English Upper Jurassic Plesiosauroidea (Reptilia) and a review of the phylogeny and classification of the Plesiosauria. *Bulletin of the British Museum (Natural History) Geology* 35:253–347.

Clark, B. D. and W. Bemis. 1979. Kinematics of swimming of penguins at the Detroit Zoo. *Journal of Zoology* 188:411–428.

Davenport, J., S. A. Munks, and P. J. Oxford. 1984. A comparison of the swimming of marine and freshwater turtles. *Proceedings of the Royal Society* B 220:447–475.

Godfrey, S. J. 1984. Plesiosaur subaqueous locomotion: a re-appraisal. *Neues Jahrbuch für Geologie und Paläontologie, Monatshefte.* 1984:661–672.

Hoar, W. S. and D. J. Randall. 1978. *Fish Physiology.* VII: *Locomotion.* New York: Academic Press.

Lang, T. G. and K. Prior. 1966. Hydrodynamic performance of porpoises (*Stenella attenuata*). *Science* 152:531–533.

Robinson, J. A. 1975. The locomotion of plesiosaurs. *Neues Jahrbuch für Geologie und Paläontologie, Abhandlungen* 149:286–332.

Romer, A. S. 1966. *Vertebrate Palaeontology.* 3d ed. Chicago: University of Chicago Press.

Romer, A. S. 1968. *The Procession of Life.* London: Weidenfeld and Nicholson.

Rothschild, B. and L. D. Martin. 1987. Avascular necrosis: occurrence in diving Cretaceous mosasaurs. *Science* 236:75–77.

Scott, P. and R. Rines. 1975. Naming the Loch Ness monster. *Nature* 258:466–467.

Wheeler, A. 1969. *The Fishes of the British Isles and North-West Europe.* London: Macmillan.

Witchell, N. 1975. *The Loch Ness Story.* Harmondsworth: Penguin Books.

X

Death of the Giant Reptiles

ALL THE magnificent animals that I have been writing about became extinct at the end of the Cretaceous period, 65 million years ago. The dinosaurs died out, leaving the birds (which seem to have evolved from them) as their only descendants. The pterosaurs, ichthyosaurs, mosasaurs, and plesiosaurs died out leaving no descendants. Many invertebrate groups became extinct including the ammonites, marine mollusks with coiled shells which are extremely common fossils in Mesozoic rocks. The extinctions were devastating, but there were also a lot of survivals. The mammals (which had all been small in the time of the dinosaurs) survived well, and so did the land plants. Most of the main groups of lizards (other than mosasaurs), snakes, turtles and crocodiles also survived. What killed the dinosaurs and left the crocodiles?

There have been a lot of suggestions about why the dinosaurs went extinct, some of them distinctly far-fetched. Two hypotheses are strongly supported at present, by different groups of scientists. One says that the earth was hit by a collossal meteorite (a lump of material from outer space) and the other says that there was a period of violent volcanic activity.

The meteorite hypothesis started with a strange observation. The latest Cretaceous rocks in central Italy and the earliest rocks of the next period (the Tertiary) are both limestones. Each contains fossil shells of foraminiferans (microscopic marine animals) typical of its period. Between these limestone layers is a layer of clay, two centimeters thick, with no fossils in it. A team of scientists led by Drs. Luis and Walter Alvarez (father and son) analyzed this clay using neutron activation analysis, a technique that can measure tiny quantities of rare elements. They found remarkably high concentrations of iridium, one of the platinum group of metals.

When I say high, I mean 9 parts per billion (9 parts in 10^9: I am using the American billion, not the larger British one). That may seem too little to get excited about, but it is 30 times higher than in the limestone immediately below or a short distance above (figure 10.1). Iridium concentrations are generally exceedingly low in the earth's crust but much higher in meteorites, typically 500 parts per billion. Could the iridium in the clay have come from a meteorite?

After the Italian rocks had been analyzed, samples of rock were taken from other places, scattered around the world, where Cretaceous and Tertiary deposits meet with no apparent interruption. These were analyzed in the same way, and high iridium concentrations were found in them all. The iridium layer seemed to be everywhere.

A meteorite hitting the earth might explode, scattering iridium-rich

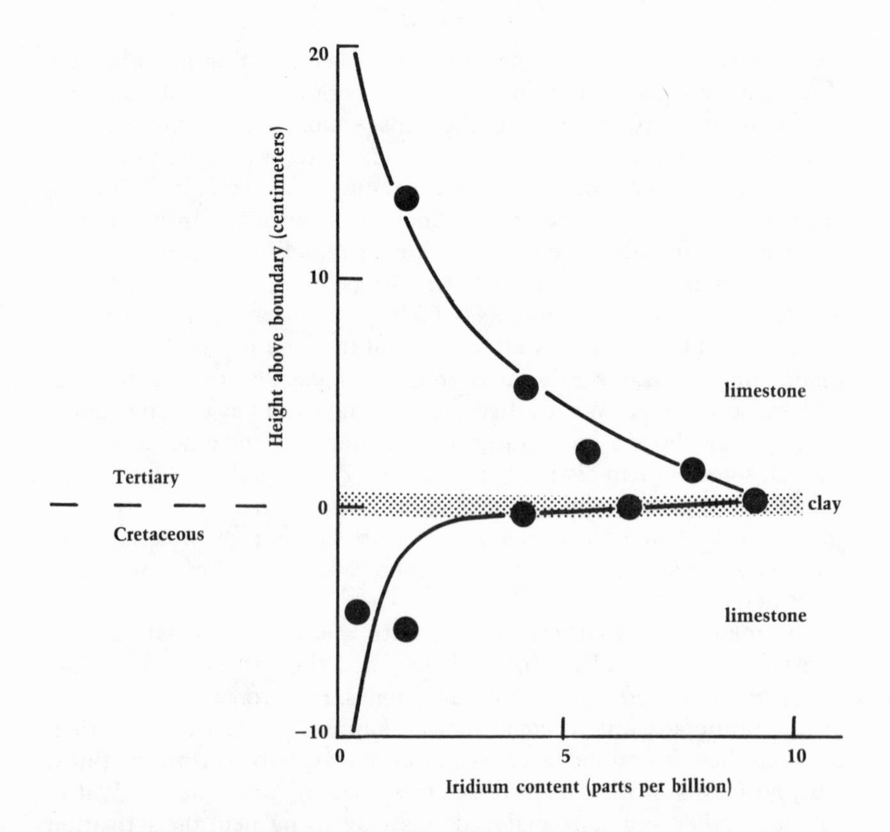

FIGURE 10.1. A graph showing iridium concentrations in the clay at the Cretaceous-Tertiary boundary at Gubbio, Italy, and in the limestones immediately above and below it. The calcium carbonate was dissolved out with acid before the analyses were carried out, and the concentrations refer to the acid-insoluble residue. The data are from L. W. Alvarez et al., *Science* (1980), 208:1095–1108.

dust, but any ordinary explosion would scatter material only over a restricted area. Here we have material scattered all over the world. The Alvarez team suggested that a huge meteorite might have disintegrated in a collossal explosion, throwing dust many kilometers up into the atmosphere. If this dust were fine enough it would be slow to settle, and might get scattered all over the earth.

It is quite easy to calculate how big the meteorite would have had to be, to scatter so much iridium. First we need to know how much extra iridium there is, above what would be expected in the same thickness of ordinary rock. Analyses from twenty-one widely scattered places give an average of 0.6 milligrams of extra iridium per square meter of the earth's surface. The area of the earth is 5×10^{14} (500 million million) square meters, so the total amount of iridium can be estimated as $0.6 \times 5 \times 10^{14} = 3 \times 10^{14}$ milligrams or 300,000 tonnes. The meteorite would probably have contained about 500 parts per billion of iridium (1 part in 2 million), so we must multiply the 300,000 tonnes by 2 million to get an estimate for the total mass of the meteorite, 600 billion tonnes.

Typical meteorites have densities of about 2.2 tonnes per cubic meter (which is rather lower than most other rocks) so a 600-billion-tonne one would have a volume of 270 billion cubic meters. A sphere of that volume would have a diameter of 9 kilometers. It would be similar in size to Manhattan Island (which is about 20 kilometers long and 4 kilometers wide).

It may seem far-fetched to imagine such a thing hitting the earth, but it is not too improbable. As well as planets orbiting the sun, there are a lot of smaller bodies called asteroids, a few kilometers in diameter. The meteorites that have been observed landing on earth seem to be fragments from collisions between asteroids, but there is a constant danger of whole asteroids hitting us. It has been estimated from telescope observations that there are about a thousand, with diameters of a kilometer or more, whose orbits take them inside the earth's orbit at times and outside it at others. None of these asteroids have collided with the earth in historic times (so far we have been lucky), but eventually some will. A few craters have been found that are believed to have been made in the distant past by small asteroids, and it has been calculated that the earth is likely to be hit by an asteroid of 10 kilometers or more diameter about once every 100 million years. That is very seldom, but the event we are trying to explain happened just once, 65 million years ago.

The earth, traveling its orbit round the sun, is hurtling through space at 30 kilometers per second (one hundred times the speed of sound in air). Asteroids travel round the sun in the same direction, so there will be no head-on collisions. Figure 10.2 shows how the earth, traveling its near-circular orbit, might be hit by an asteroid with a more elliptical orbit. Astronomers tell us that the speed of an approaching asteroid, *relative to the earth,* would probably be something like 20 kilometers per second.

We will calculate the energy of an impact at this speed. This is the kinetic energy of the asteroid, due to its movement relative to the earth. Kinetic energy is $\frac{1}{2}$ (mass) × (speed)^2, but in using the formula we must be careful about units. Six hundred billion tonnes is 600 million million kilograms. Twenty kilometers per second is 20,000 meters per second. If we put into the formula the mass in kilograms and the speed in meters per second we get the energy in joules; it is about 10^{23} (100,000 million million million) joules. This is enormous—equivalent to the explosion of 60 million megatonnes of TNT. The atomic bombs dropped in Japan had energies of only 0.02 megatonnes each. The biggest explosion of modern times, the explosion of the island volcano Krakatoa in 1883, had about 200 megatonnes energy.

My calculation is very rough because the speed that I used in it could be badly wrong, but it seems clear that the landing of the asteroid would have been incomparably more devastating than anything people have ever experienced. The energy would have been amply sufficient to vaporize the asteroid. Twenty million joules are needed to vaporize a kilogram of rock so 600 million million kilograms could be vaporized by

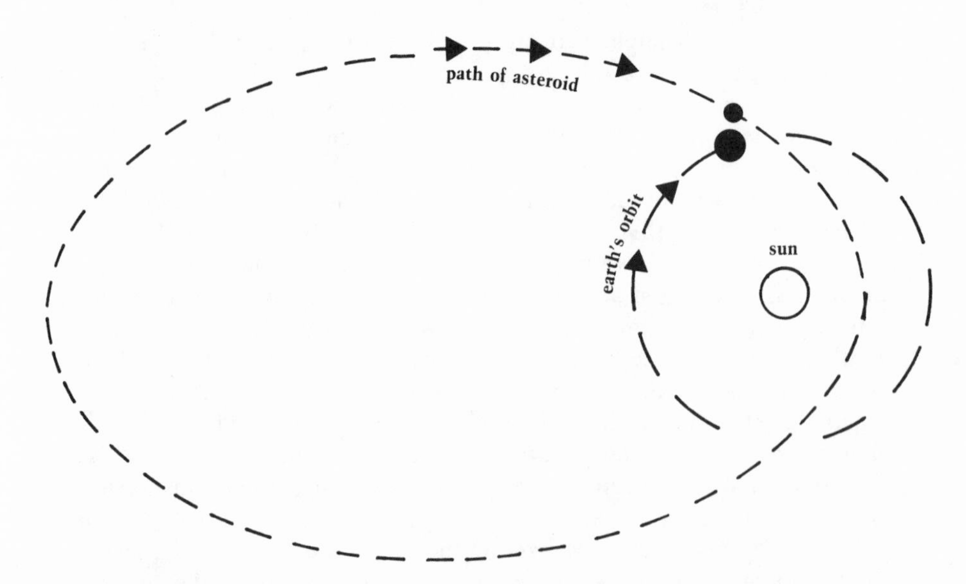

FIGURE 10.2. Collision between the earth and an asteroid. Not to scale.

12,000 million million million joules (1.2×10^{22} joules). This is only one-eighth of the estimated kinetic energy.

If the asteroid had hit dry land it would have made an enormous hole, but no crater that could have been made by it has been found. This need not worry us: it would be more likely to land in the oceans that cover 70 percent of the earth's surface. If it did fall in an ocean it would make less of a crater, and even a big crater in the ocean floor would be hard to find.

A lot of the energy of the asteroid would be transmitted to the ocean water, turning some of it to high-pressure steam, but there would be enough left to vaporize most of the meteorite. There would be a terrific explosion that would blast a column of steam and rock vapour high into the atmosphere where they would condense out as tiny ice crystals and dust particles. The dust would sink down again onto the earth but the rate of settling would depend on the size of the particles: it would probably take a few months.

Now we will think about how the catastrophe could have affected dinosaurs and other animals. First, the dust in the atmosphere would have blotted out the sun. Sunlight all over the world is dimmed by dust after major volcanic eruptions, and would be dimmed far more by the catastrophe we are imagining. It has been estimated that after the Krakatoa eruption, sunlight was dimmed by 3 percent to 0.97 of its normal intensity. Twice as much dust would dim it to $(0.97)^2$ times the usual intensity, three times as much to $(0.97)^3$ and so on. If the asteroid threw up 200 times as much dust as Krakatoa (and the ratio of energies suggests it would have thrown up far more than that) sunlight would be dimmed to $(0.97)^{200} \simeq 0.002$ times its normal intensity: the world would have been plunged in darkness.

Calculations of the size of the dust particles and of the rate at which they would settle suggest that the darkness would have lasted for several months. Plants would have suffered because they depend on the energy of sunlight to make foodstuffs by the process of photosynthesis. Many land plants would probably have survived a few months darkness (remember that deciduous trees lose their leaves each fall and have to survive a few months without photosynthesis), and other land plants would probably have survived as seeds. However, the microscopic plants that float as plankton in lakes and seas would probably have suffered badly, because they are too small to have substantial food reserves. Almost all the animals in lakes and seas get their energy ultimately from these microscopic plants: tiny crustaceans and other animal plankton eat the plants and are in turn eaten by fish and other larger animals. Life in lakes and seas would suffer very badly.

A second effect would result from the blotting out of sunlight: the earth's surface would get cold. The effect on the oceans would not be great because of their enormous heat capacity, but there would be severe frosts on land that might be lethal to many plants and animals.

A third possible effect would be acid rain. The high temperatures of the explosion would make some of the nitrogen in the atmosphere combine with oxygen to form nitrogen oxides. These would react with water and more oxygen to form nitric acid, which would fall from the atmosphere in rain.

Acid rain due to very different causes is a serious modern problem. Nitrogen oxides are formed in the engines of motor vehicles and released into the atmosphere with the exhaust. Sulphur dioxide is emitted by coal-burning power stations. The nitrogen and sulphur oxides react with oxygen and water to form nitric and sulphuric acids, which fall in rain. The acid rain falling on trees makes leaves yellow and fall off. When it drains into lakes it makes them acid, sometimes too acid for fish to survive. Many forests in industrial countries are in poor health and many lakes have lost their stocks of fish. The acid rain after the asteroid explosion would have had similar effects.

The asteroid explosion would have had disastrous effects on many kinds of animals and plants. Nevertheless, the asteroid hypothesis for the extinctions at the end of the Cretaceous has several problems. One is that the high iridium concentrations are not limited to very thin layers of rock as would be expected if they had been formed by dust settling in the few months after the explosion. Instead, they extend through thicknesses of 30 to 100 centimeters, that probably took several tens of thousands of years to form. Indeed, some American samples show several iridium-rich layers sandwiched between iridium-poor ones. Another problem is that the extinctions do not seem to have happened all at once. The numbers of ammonite and dinosaur species seem to have declined gradually during the last few million years of the Cretaceous, and the last North American dinosaurs are in rocks above the iridium-rich layer, formed 40,000 years after it.

These observations seem to favor the volcanic hypothesis, which postulates a few tens of thousands of years of intense volcanic activity. Volcanoes throw up material from deep inside the earth, where iridium concentrations are much higher than in surface rocks, though lower than in meteorites. The iridium-rich layers may have come from volcanoes.

Major eruptions throw enough dust into the upper atmosphere to dim sunlight perceptibly, but eruptions big enough to black out the sun and stop photosynthesis seem unlikely. Volcanoes seem more likely to cause extinctions by way of acid rain. They emit sulphur dioxide and other gases as well as molten rock. The sulphur dioxide (like the sulphur dioxide from power stations) eventually falls in rain as sulphuric acid. Scientists have analyzed emissions from a Hawaiian volcano, measuring the quantities of sulphur dioxide and iridium. Their measurements suggest that if the 300,000 tonnes of iridium in the iridium-rich layers came from volcanoes, they would have been accompanied by about 10 million million tonnes of sulphur dioxide. If this was emitted over a period of 10,000 to 100,000 years, the average rate of sulphur dioxide emission would have been between 100 million and 1 billion tonnes per year. The rate would probably vary, with peak rates far above average.

The surphur dioxide, released annually in the United States and Europe by burning coal and oil, totals 80 million tonnes, and the nitrogen oxide emissions are less. I have already described the damage that these emissions are causing, through acid rain. The acid rain in the worst parts of the supposed period of volcanic activity would have been far worse.

It would have been so much worse that the oceans would have been seriously affected, as well as lakes. At present the oceans are slightly alkaline with a pH value of 8.2, but the acid from 10 million million tonnes of sulphur dioxide would reduce the pH to about 7.4. This is still very slightly on the alkaline side of neutrality (pH7 is neutral and anything less is acid), but it is not alkaline enough for foraminiferans. It is also near the limit for another group of microscopic plankton, the coccoliths.

Foraminiferans have calcium carbonate shells and coccoliths have plates of calcium carbonate on their outer surfaces, and calcium carbonate dissolves in acid. Foraminiferans need a pH of 7.6 or more and coccoliths need at least 7.0–7.3. Both groups suffered many extinctions at the end of the Cretaceous but dinoflagellates and other groups of plankton, without calcium carbonate skeletons, survived better.

Ozone is a form of oxygen that is rare at ground level but relatively more plentiful in a layer high in the atmosphere. It absorbs much of the ultraviolet radiation from the sun, protecting living things from these harmful rays. The ozone layer is depleted after major eruptions because volcanoes inject a little hydrochloric acid into the atmosphere, as well as the much larger quantity of sulphuric acid. The hydrochloric acid reacts with the ozone in the ozone layer to form chlorine, water, and ordinary oxygen. It has been calculated that 8 percent of the ozone was destroyed after the Krakatoa eruption. The many eruptions that are supposed to have happened at the end of the Cretaceous would have nearly destroyed the ozone layer, leaving animals and plants exposed to abnormally high doses of ultraviolet radiation. This might have been fatal for many of them, but the mammals of the time were small and would have survived if they spent their days in burrows and were active mainly at night.

The volcanic hypothesis seems quite attractive but there is at least one observation that it seems unable to explain. Damaged sand grains have been found wherever they have been looked for, in samples from the iridium-rich layer from various places, scattered around the world. Sand grains are small quartz crystals, made of neatly stacked layers of atoms. The damaged ones look cracked (when examined under the microscope) because some of their layers of atoms have been thrown into disarray. Similar damage is found in sand grains from meteorite craters and is believed to have been caused by shock waves. The volcanic hypothesis seems unable to explain shocked quartz being scattered widely round the world.

Both hypotheses seem reasonably plausible. Collision of an asteroid with the earth, and a prolonged period of fierce volcanic activity, would each have had dire consequences, and would probably have caused widespread extinctions. The volcanic hypothesis is possibly the better of the two, in explaining why the extinctions were so selective: acid rain would kill foraminiferans but not dinoflagellates, and ultraviolet radiation would be more damaging for diurnal dinosaurs than for nocturnal mammals. However, the asteroid hypothesis seems better able to explain the shocked quartz. The supporters of the two hypotheses are still arguing fiercely, and it is quite possible that neither is right. Indeed, it has recently been suggested that the extinctions were due to another cause, a shower of comets hitting the earth. The effects of comet impacts would be much like those of asteroid impacts: comets are massive bodies travelling at exceedingly high speeds, and contain more iridium than the surface rocks of the earth. A series of comet impacts could explain extinctions spaced out in time, and there are theoretical reasons for expecting comets to come in showers lasting about a million years. The implications of the comet shower hypothesis have not yet been worked out in as much detail as those of the asteroid and volcano hypotheses. It remains to be seen which (if any) of the three hypotheses triumphs.

Principal Sources

The meteorite hypotheses was put forward by Alvarez et al. (1980) and discussed in much more detail by Silver and Schultz (1982). The volcano hypothesis has been presented by Officer et al. (1987). Sloan et al. (1986) published evidence that the dinosaurs declined gradually. Bohor, Modreski, and Foord (1987) presented the shocked quartz evidence. Hut and others (1987) presented the comet shower hypothesis.

Alvarez, L. W., W. Alvarez, F. Asaro, and H. V. Michel. 1980. Extraterrestrial cause for the Cretaceous-Tertiary extinction. *Science* 208:1095–1108.

Bohor, B. F., P. J. Modreski, and E. E. Foord. 1987. Shocked quartz in the Cretaceous-Tertiary boundary clays: Evidence for a global distribution. *Science* 236:705–709.

Hut, P. and others (1987). Comet showers as a cause of mass extinctions. *Nature* 329:118–126.

Officer, C. B., A. Hallam, C. L. Drake, and J. D. Devine. 1987. Late Cretaceous and paroxysmal Cretaceous/Tertiary extinctions. *Nature* 326:143–149.

Silver, L. T. and P. H. Schultz. eds. 1982. Geological implications of impacts of large asteroids and comets on the earth. *Geological Society of America Special Paper* 190:1–528.

Sloan, R. E., J. K. Rigby, L. M. Van Valen, and D. Gabriel. 1986. Gradual dinosaur extinction and simultaneous ungulate radiation in the Hell Creek Formation. *Science* 232:629–633.

XI

Giant Birds

ARCHAEOPTERYX, THE oldest known bird, lived in the Jurassic period, quite early in the time of the dinosaurs. It was only about the size of a magpie and no giant birds appeared until the Tertiary, after the dinosaurs had gone.

The teratorns were birds of prey that lived in America then. Figure 11.1 will give you an impression of their size: (b) shows the wing bones of a Californian condor, one of the largest modern birds of prey; (d) shows the wing bones of the Wandering albatross, which has the biggest wing span of all modern birds; (c) shows the wing skeleton of the best-known teratorn, *Teratornis merriami:* it is almost as long as the albatross wing skeleton, and rather stouter; finally, (a) is amazing. It is the humerous (the bone from the base of the wing) of *Argentavis magnificens,* the biggest known teratorn. It is more than twice as long as the condor humerus shown immediately below it.

Only one skeleton of *Argentavis* has been found (in Argentina). The wing bones are incomplete but the pieces that have been found are enough to show that the wings were strong and well developed. It seems impossible to avoid the conclusion that those huge wings were for use, that *Argentavis* could fly although it was far bigger than any modern flying bird.

The wing span of the condor is ten times the length of the humerus. If *Argentavis* was built to the same proportions its span was about 6 meters, far more than the spans of the condor (2.7 meters) or albatross (3.4 meters). Its biggest wing feathers must have been about 1.5 meters long and 200 millimeters wide. Its body mass has been estimated from the circumference of a leg bone in the same way as was done for dinosaurs, using a graph like Figure 2.5 showing tibiotarsus circumference and body mass for 324 species of modern birds. The body masses corresponding to the tibiotarsus circumferences of teratorns were read

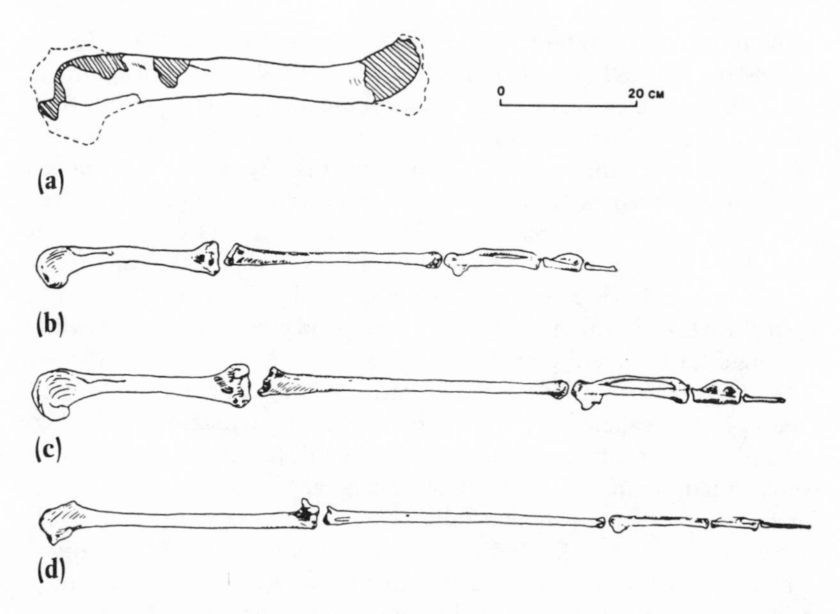

FIGURE 11.1. Wing bones of (a) *Argentavis*, the largest teratorn; (b) a Californian condor; (c) a Wandering albatross; and (d) *Teratornis*. From Campbell and Tonni 1983.

off from the graph: 14 kilograms for *Teratornis merriami* and 80 kilograms (heavier than most men) for *Argentavis*.

That 80 kilograms is the best estimate that has been made of the mass of *Argentavis*, but there is a lot of uncertainty about it. Birds of equal mass, of different species, may have considerably different bone thicknesses. Statistical analysis of the data tells us that the conclusion should be no more precise than this: there is 95 percent probability (the odds are 19 to 1) that the mass of *Argentavis* lay between 37 and 166 kilograms. Even with that much uncertainty it is clear that *Argentavis* was *much* heavier than the Californian condor (about 10 kilograms) or even the Kori bustard, which reaches about 16 kilograms and seems to be the heaviest modern flying bird.

Argentavis was also much heavier than *Pteranodon*, the giant pterosaur (figure 8.2), which was so lightly built that its mass seems to have been no more than about 15 kilograms. However, *Pteranodon* had the larger wing span (7 meters. We estimated the span of *Argentavis* as only 6 meters). The few fragments that have been found of the even larger pterosaur, *Quetzalcoatlus*, suggest a span of about 12 meters and a mass of possibly about 60 kilograms. *Argentavis* and *Quetzalcoatlus* are rivals for the title of the biggest flying animal of all time.

The largest modern birds cannot fly . They are the ostrich (up to 120 kg), cassowaries (60 kg), the emu (50 kg) and the Emperor penguin (40 kg). With the exception of the penguin, these are members of the group called the ratites, which also includes the rheas and kiwis.

Ostriches and other ratites are like enormously overgrown chicks. They have tiny wings, useless for flight, and well-developed legs. They have fluffy plumage instead of the blade-like feathers of other adult birds. They also have some chick-like features in their skeletons. They are believed to have evolved by processes of development getting out of step with each other: they grow large and sexually mature while keeping a lot of juvenile features.

The biggest extinct birds are also ratites. They are the moas, which lived in New Zealand, and the elephant birds, in Madagascar. The biggest moa is *Dinornis maximus,* 3.5 meters tall (twice the height of an average man, figure 11.2). The biggest elephant birds looked very similar and were about 3 meters tall. There were also some giant birds that were not ratites. *Diatryma,* a wicked-looking predator that lived in North America, was about 2 meters tall (figure 11.2). It lived quite early in the Cenozoic era but the moas and elephant birds are more recent. Indeed, the moas survived in New Zealand until after the Maoris arrived.

I have made scale models of moas and used them to estimate the

FIGURE 11.2. *Dinornis maximus* (the largest moa), *Diatryma steini,* and an adult man, all to the same scale.

masses of the living birds, in much the same way as I estimated the masses of dinosaurs (chapger 2). I modeled the main features of the skeleton in wire and then added clay to represent the flesh, making models that represented the birds as if they had been plucked. I measured the volumes of the models and used them to calculate the masses of the plucked birds, assuming that their densities were the same as the density of a plucked goose carcase (which I measured). Finally, I added an allowance for the feathers, which I assumed to be the same fraction of body mass as in turkeys and kiwis. The result for a big *Dinornis* was 240 kilograms, about the same as a large tiger. The biggest elephant birds were stouter, although they were a little shorter, and may have been nearly twice as heavy.

Though *Dinornis* was the biggest moa, the one that fascinates me is *Pachyornis elephantopus,* shown in figure 11.3. Its splendid name means "fat bird with elephant's feet," and seems very suitable. Its leg bones are amazingly thick. Compare it with the ostrich, drawn beside it to the same scale. I estimate this moa's mass as 130 kilograms (from measurements on a model) and the ostrich, whose skeleton is illustrated, as only 68 kilograms, but even so the moa bones look disproportionately thick.

Appearances can be deceptive, so I measured the moa's bones and calculated strength indicators in the same way as for dinosaurs (chapter 4). The results are shown in table 11.1. The value for the tibiotarsus (shin bone) is about the same as for an ostrich, and those for the other two bones are twice as high as for the ostrich. To be consistent with my line of argument in chapter 4, I should conclude that *Pachyornis* was at least as athletic as ostriches, but I find that hard to believe. Ostriches are exceedingly fast runners, probably faster than the African antelopes. Should I conclude that moas were also exceedingly fast? I cannot believe that they were, with those clumsy-looking legs.

The key to the problem may be that moas seem to have had no need to run. They fed on plants, as remains of their stomach contents show, and so had no need to run to capture food. There seem to have been no big predators in New Zealand, until the Maoris arrived, so there was nothing to run away from. (That is to say, there was nothing to run away from while they were evolving. They seem to have been easy prey to the Maoris, who hunted them to extinction.) It was not the same for the dinosaurs: the flesh-eating dinosaurs had to run to catch prey and the plant-eating ones had to run to escape. This difference between moas and dinosaurs may justify a different interpretation of their strength indicators.

My idea involves safety factors. Suppose an engineer is designing a small bridge to carry a maximum load of ten tonnes. He would be an optimist if he calculated the thickness of steel that could just support ten tonnes without breaking, and ordered steel that thick. Any reputable engineer would allow a safety factor: he might design the bridge to be able to carry twenty tonnes although he expected the maximum load to be only ten. The reason for this is that neither load nor strength can be predicted precisely. An unexpectedly large load may arrive, or the steel may be substandard, and in either case a bridge that was expected to be strong enough may fail. The bigger the safety factor the less likely this is to happen.

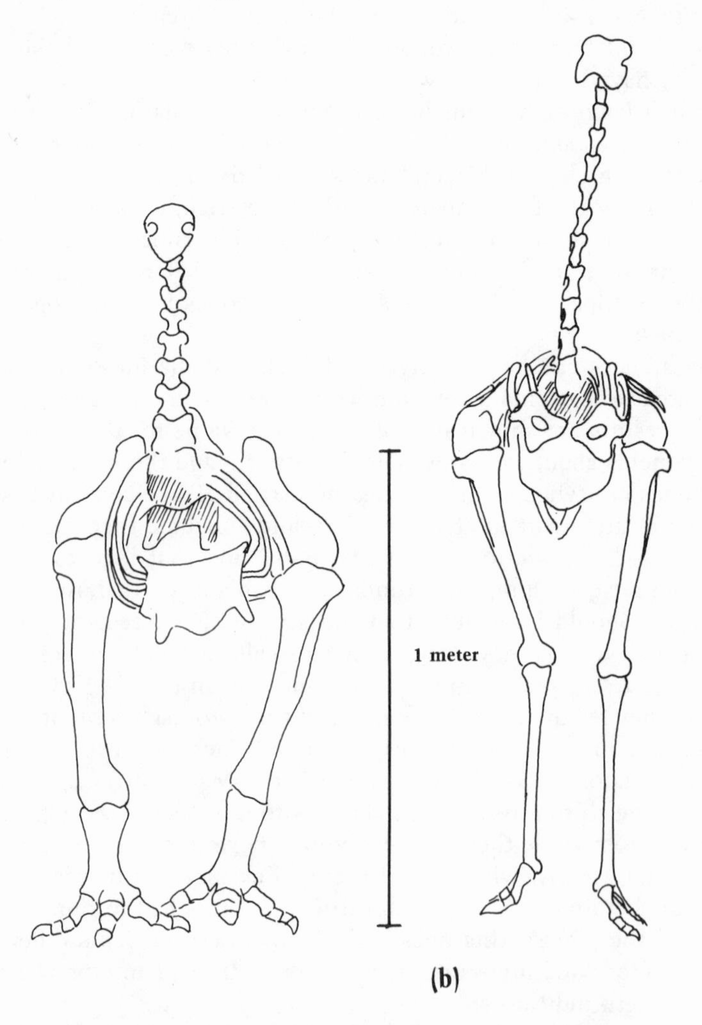

FIGURE 11.3. Skeletons of (a) *Pachyornis* and (b) an ostrich, to the same scale. From Alexander 1983a.

TABLE 11.1. Strength indicators ($Z/W\chi$, see p. 53) for leg bones of an ostrich and the moa *Pachyornis elephantopus.*

	Strength indicator (square meters per giganewton) for:		
	Femur	*Tibiotarsus*	*Tarsometatarsus*
ostrich	45	18	17
moa	94	17	39

A stronger bridge is less likely to fail but costs more to build. Beyond a certain point, the slight advantage of extra safety obtained by making it stronger still is not worth the extra cost. The ideal strength depends on the cost of the materials. If we had to make bridges of platinum we would make them weaker and live more dangerously. If cheap second-hand steel were available we might make a bridge extra strong.

The evolution of skeletons has also involved balancing safety against cost. In this case the cost is partly the cost in energy and materials of growing a stronger bone, but is largely the penalty of having to carry extra bone around. Thick leg bones may be less likely to break in a fall, but they make it harder to run fast, just as people find it hard to sprint in heavy boots. Measurements and calculations on ostriches and several kinds of mammal have shown that their leg bones are about three times as strong as is necessary to withstand the forces involved in strenuous activities such as running and jumping. These bones are built to safety factors of about three, and so can stand many of the larger forces that occur accidentally, for example in falls and collisions.

Moas may have had larger safety factors. However strong their bones, there would always be some danger of an accident bad enough to break them. If they had no need to run, they might not be inconvenienced much by heavy bones. The cost of extra strength might be less for them than for ostriches, which have to run to escape from lions. If strength were cheap, ideal safety factors would be high, which may explain the remarkably thick leg bones of *Pachyornis* and (to a lesser extent) other moas.

Several sets of moa footprints have been found and I have calculated speeds for them, in the same way as for dinosaurs (chapter 3). All of them seem to have been moving between 0.8 and 2.0 meters per second, which would have been walking speeds. This does not prove that they never ran, but at least it does not contradict the suggestion that they were not very athletic.

My belief, that moa leg bones had high safety factors, has been challenged. Palaeontologists in Tübingen have suggested that moas may have lived in thick undergrowth and may have needed very strong legs to force their way through. I find that hard to believe. An animal that behaved like a bulldozer would use a lot more energy than one that slipped through small gaps, or avoided the densest patches of vegetation, and might be a poor competitor. Nevertheless, the possibility should be considered.

The legs of elephant birds are little less remarkable than the legs of moas, but it is their eggs that I want to write about, the biggest of all known eggs. Quite a lot of elephant-bird egg shells have been found in Madagascar, some with the bones of embryos still inside them. The eggs of the biggest species are 30 centimeters long with a volume of 9 liters (2.4 U.S. gallons). Ostrich eggs are only about half as long, with a volume of 1.3 liters, and the eggs of all modern reptiles are much smaller. Even known dinosaur eggs are smaller than elephant bird eggs: the biggest I know of are only 25 centimeters long.

Let us think what problems there might be, for very large eggs. First there is the problem of ventilation. Bird embryos are not hermetically sealed in their eggs, like cans of soup. The eggshell is porous, allowing gases to diffuse in and out. This enables the embryo to get the oxygen it needs for respiration, and to get rid of waste carbon dioxide.

Think of two eggs, one twice the length of the other. It has eight times the volume of the smaller egg, and the embryo in it, just before hatching (when it uses oxygen fastest), is eight times as heavy. The big embryo uses oxygen faster than the small one, but not eight times as fast, because rates of oxygen consumption are not strictly proportional to body mass either for adult animals (figure 7.1) or for embryos. The large embryo will probably use oxygen only four or five times as fast as the small one.

The more pores there are, or the wider the pores, the faster oxygen can diffuse in. However, the thicker the shell, the further the oxygen has to diffuse and the lower the rate of diffusion. The maximum rate of diffusion that a shell allows is proportional to

$$\frac{\text{number of pores} \times \text{area of each pore}}{\text{thickness of shell.}}$$

If the two eggs were precise scale models of each other they would have equal numbers of pores and the larger one would have pores of twice the diameter, therefore four times the cross-sectional area, but its shell would be twice as thick. It would allow oxygen to diffuse just twice as fast but the embryo in it would need oxygen, as we have seen, four or five times as fast. This tells us that big eggs need more porous shells than small ones. An excessively big egg would need a shell so riddled with pores as to be seriously weakened. If this shell were made thicker, to strengthen it, it would have to be more porous still. The need to be sufficiently porous must set an upper limit to the sizes of eggs.

Even elephant bird eggs are probably a long way from that limit. Chicken eggs have very sparse pores, piercing only 0.02 percent of the are of the shell. Ostrich eggs have to be much more porous, with 0.2 percent of their area accounted for by pores. Elephant bird eggs must have been more porous still, but even if the pores were 2 percent of their area the shells would not be seriously weakened. I know no measurements of their pores so I cannot state the exact percentage.

Elephant bird eggs may be near an upper size limit, for a different reason. Eggs have to be strong enough to withstand the forces that act on them, when the parent birds get on and off the nest, but they must be weak enough for the hatching chick to break its way out. Think again of two eggs, one twice as long as the other and eight times as heavy. It probably needs to be more than eight times as strong. This is because big birds are heavier, relative to the masses of their eggs, than small ones: very small birds are about five times as heavy as their eggs but ostriches are fifty times as heavy as their eggs. The big egg seems to need to be more than eight times as strong, but if its shell is just twice as thick, it will only be four times as strong. (It is a general rule for objects of the same shape, made of the same material, that strength is proportional to $(\text{length})^2$). This means that bigger eggs need relatively thicker shells. An egg that is twice as long as another is generally found to have a shell about three times as thick. The eggshell is 4 percent of the mass of a hummingbird egg but 17 percent of the mass of an ostrich egg.

Hatching chicks break their eggs open by hammering at the shell. When you break something by hammering or by any other kind of impact, what decides whether it breaks or not is the energy of the blow. A heavy hammer brought down fast has more kinetic energy than a light hammer moving slowly, and is more likely to break things. Theory tells us that the energies needed to break egg shells should be about proportional to the masses of the shells. The energies that chicks can put into blows should be about proportional to the masses of the chicks and so to the masses of the egg contents (since the hatching chick fills the shell). If bigger eggs have relatively thicker shells, it will be harder for their chicks to break out of them. Too big an egg would be an unbreakable prison.

One final thought about elephant birds: they must have been blessed with patience. Small egg hatch soon but big ones take longer. Very small bird eggs hatch in about 15 days, and ostrich and emu eggs take about 50 days. If the trend continues, elephant bird eggs would have taken about 90 days to hatch.

This chapter has been about three groups of giant birds: the teratorns, moas, and elephant birds. The teratorns had well-developed wings and could presumably fly, though the biggest seem to have been five times as heavy as any modern flying bird.

The moas and elephant birds were like outsize ostriches and could not fly. Some moas had astonishingly thick leg bones which seem unnecessarily strong, for animals that do not look very athletic. I suggest that they may have evolved unusually high safety factors because extra mass in the legs would be little disadvantage, if moas did not have to run. They were not threatened by any predator until humans arrived in New Zealand.

Elephant birds laid the biggest known eggs. These needed thick shells to protect them from damage by the parents, but thick shells need to be very porous (to let oxygen diffuse in fast enough) and are difficult for the hatching chick to break out of. Elephant bird eggs may have been near the limit of size set by the difficulty of hatching.

Principal Sources

Campbell and Tonni (1983) discussed the flying ability of teratorns. The discussion of moas is based on two papers of my own, one (1983) on *Pachyornis* and one (1981) on safety factors in animal skeletons generally. Most of my information about eggshells comes from Anderson, Rahn, and Prange (1979), Rahn, Ar, and Paganelli (1979), and Tullett (1984).

Alexander, R. McN. 1981. Factors of safety in the structure of animals. *Science Progress* 67:109-130.

Alexander, R. McN. 1983a. Allometry of the leg bones of moas (Dinornithes) and other birds. *Journal of Zoology* 200:215–231.

Alexander, R. McN. 1983b. On the massive legs of a moa (*Pachyornis elephantopus*, Dinornithes). *Journal of Zoology* 201:363:376.

Anderson, J. F., H. Rahn, and H. D. Prange. 1979. Scaling of supportive tissue mass. *Quarterly Review of Biology* 54:139-148.

Campbell, K. E. and E. P. Tonni. 1983. Size and locomotion in teratorns (Aves: Teratornithidae). *The Auk* 100:390-403.

Rahn, H., A. Ar, and C. V. Paganelli. 1979. How bird eggs breathe. *Scientific American* 240(2):38-47.

Tullett, S. G. 1984. The porosity of avian eggshells. *Comparative Biochemistry and Physiology* 78A:5-13.

Reif, W. E. and H. Sylin-Roberts. 1987. On the robustness of moa leg bones. *Neues Jahrbuch für Geologie and Paläontologie, Monatshefte* 1987:155-160.

XII

Giant Mammals

THERE WERE no big mammals while the dinosaurs lived but many mammals of elephant or rhinoceros size evolved during the Cenozoic era.

The best known of the giant extinct mammals were the mastodons and mammoths. Mastodons are primitive elephants, distinguished by their simple teeth. Mammoths are much more like modern elephants. Both survived to overlap in time with humans, and many cave paintings of mammoths have been found.

The biggest mammoth species (*Mammuthus imperator* of North America) stood 4 meters tall at the shoulder. Very large (6 tonne) African elephants are only 3.3 meters tall. The mammouth, 1.2 times as tall, must have been about 1.2^3 times as heavy, about 10 tonnes. Mastodons were less tall but had relatively longer bodies.

Modern elephants live in hot places where they seem to have trouble keeping cool, but many mammoths lived in temperate or even cold places. Cave paintings in France and Spain show mammoths with long hair, which they may have needed for warmth during the Ice Ages. Frozen carcases of *Mammuthus primigenius* have been found embedded in ice in Siberia: the mammoths seem to have fallen into crevasses, died and frozen, and to have remained frozen until they were found. They have long black hair, just as shown in the cave paintings, and they also have an 8-centimeter layer of fat under the skin. Both the hair and the fat may have been useful as heat insulation. Modern elephants can maintain their body temperatures in the warm climates of Africa and India, and also in zoos in temperate countries, but mammoths inhabited much colder environments and probably needed extra insulation.

Most extinct mammals, including the mammoths, seem unspectacular in comparison with dinosaurs. The saber-tooths (*Smilodon*), which preyed on mammoths, were impressive, but they were only about the size of modern lions. The giant Irish deer (*Megaloceros*) was smaller in the body than a moose though its antlers grew to a span of 3.5 meters. The giant ground sloth (*Megatherium*) of South America was enormously larger than any modern sloth but its estimated mass (3 tonnes) is no more than that of the biggest modern rhinoceros.

There are various other extinct mammals of modern rhinoceros size, and just one that is enormously larger. It is *Indricotherium* (formerly called *Baluchitherium*), a hornless rhinoceros from Mongolia. It is hard to be sure of its maximum size because no complete skeleton has been found, only odd bones from specimens of various sizes. Only two neck vertebrae and a foot bone (a metacarpal) seem to come from the biggest size of skeleton. These three bones have been drawn to scale in figure 12.1, and the rest have been scaled up from smaller skeletons. The sizes of some ribs and vertebrae have had to be guessed because no specimens were found of those particular bones. Thus the evidence for the size of the animal in figure 12.1 is shaky: a small error of judgment could have made it badly wrong. Since no better evidence is available, I will assume that the figure is accurate. It shows an animal 5.3 meters tall, much taller than any elephant.

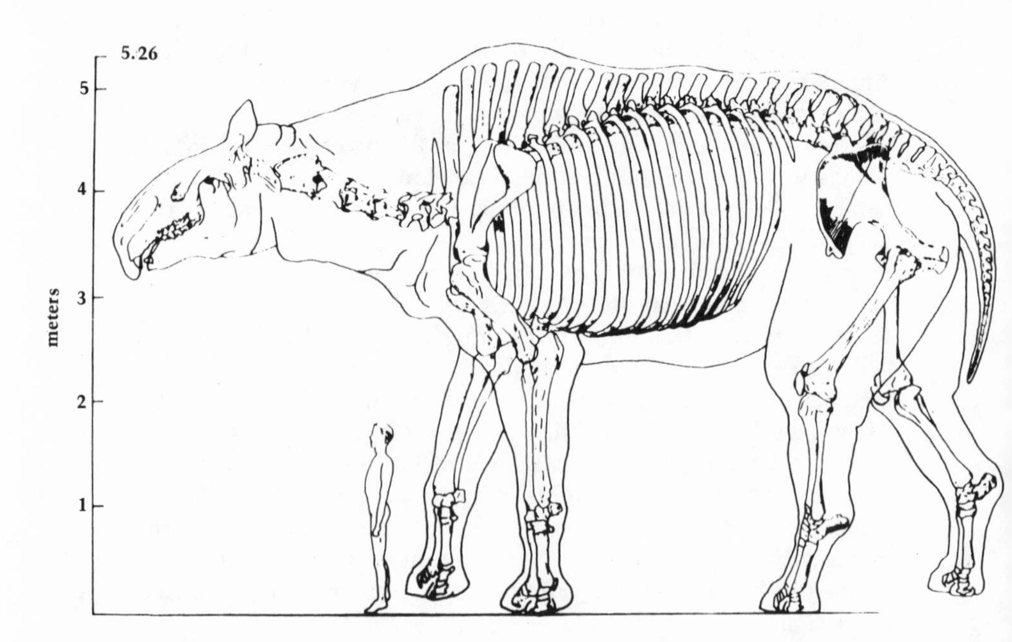

FIGURE 12.1. A reconstruction of the largest *Indricotherium*. From Granger and Gregory (1935).

The scientists responsible for the picture estimated the mass of the animal to be 20 tonnes, but I think it may have been even more. The head and body (excluding the tail) are 9.2 meters long, measured along the curve of the back. The same measurement in 0.75-tonne African buffalo is 2.6 meters. The *Indricotherium* has a body of roughly buffalo-like shape, so if it ws (9.2/2.6) times as long as the buffalo it was $(9.2/2.6)^3$ times as heavy: about 34 tonnes. I have tried calculating its mass in other ways and obtained even larger estimates. If the restoration is accurate, *Indricotherium* had about the same mass as *Apatosaurus* (figure 1.7).

When we discussed the heat balance of large dinosaurs (chapter 7) we were uncertain whether they had reptile-like or mammal-like metabolism. We concluded that a large sauropod with mammal-like metabolism would have to evaporate a lot of water to avoid overheating in hot climates. *Indricotherium* is obviously a mammal and presumably had mammal-like metabolism, but the climate in Mongolia must have been reasonably cool in its time, as it is now. The continents had by then reached their present positions on the earth's surface.

Indricotherium is the only land-living mammal known to have grown to the size of large sauropods, but there are bigger mammals in the sea. The Blue whale (*Balaeonoptera musculus*) grows to an average adult mass of about 100 tonnes and is the biggest animal known to have lived, at any time. Its numbers have been very seriously reduced by whaling but, happily, it survives, so it needs no further discussion in this book on extinct giants.

Principal Sources

Savage and Long (1986) describe the extinct groups of mammals. Economos (1981) discusses *Indricotherium.* Granger and Gregory (1936) is the source for figure 12.1.

Economos, A. C. 1981. The largest land mammal. *Journal of Theoretical Biology* 89:211-215.

Granger, W. and W. K. Gregory. 1935. A revised restoration of the skeleton of *Baluchitherium,* gigantic fossil rhinoceros of Central Asis. *American Museum Novitates* 787:1-3.

Savage, R. J. G. and M. R. Long. 1986. *Mammal Evolution: An Illustrated Guide.* London: British Museum (Natural History).

XIII

Epilogue

THIS BOOK about gigantic animals has highlighted some of the special problems of large size, many of which depend on the rule of squares and cubes. The weights of geometrically similar animals of different sizes are proportional to the cubes of their lengths, but the areas of corresponding body surfaces are proportional only to the square of length: an animal twice as long as another animal of the same shape is eight times as heavy but has only four times the area. This is why large animals get bogged down in mud more easily than small ones (chapter 3): their weights are proportional to the cube of length but the areas of the soles of their feet only to the square. It is why large flying animals must fly fast, to keep themselves airborne, and may have trouble taking off (chapter 8): their weights are proportional to the cube of length but the areas of their wings only to the square. It is also why large animals are not as athletic as small ones (chapter 4): the forces that act on them in dynamically similar activities are proportional to body weight and so to the cube of length but the strengths of bones and muscles (which depend on cross-sectional area) only to the square. This last example is a little more complicated than the others because the transverse force that a bone can stand, acting on its end, is not directly proportional to area but to the ratio Z/x: however, Z/x is proportional to the square of length in geometrically similar animals.

We depart further from the simple rule of squares and cubes in questions of heat balance (chapter 7). Rates of metabolic heat production are not proportional to body mass but more nearly to $(\text{body mass})^{0.75}$. Rates of loss of heat, for equal temperature differences between body and environment, are not simply proportional to surface area but depend also on the thickness of skin and any additional insulating layer. However, the effect of the relationships is that large animals can be more effectively endothermic ("warm-blooded") than small ones. Shrews and other small endotherms need relatively thick fur or feathers but large ones such as elephants need no fur at all and very large endotherms might be liable to overheat. Similar conclusions apply to eggshells (chapter 11) because the mathematics of gas diffusion resembles the mathematics of heat conduction: big eggs (such as those of ostriches) need more porous shells than small ones (such as those of sparrows) to allow gases to diffuse in and out fast enough to sustain their metabolism. Finally, large ectothermic ("cold-blooded") animals take longer than small ones to equilibrate to a new environmental temperatures. Small lizards approach equilibrium within a few minutes but a large ectothermic dinosaur would take many days.

Those arguments depend on assumptions of geometric similarity, but large animals are not geometrically similar to small ones. Elephants are not the same shape as shrews, nor are albatrosses the same shape as hummingbirds. Masses of geometrically similar animals would be proportional to $(\text{bone circumference})^3$ and the masses of real mammals are proportional to $(\text{bone circumference})^{2.73}$ (figure 2.5). Wing loadings of geometrically similar birds would be proportional to $(\text{body mass})^{0.33}$ and actual wing loadings are proportional to $(\text{body mass})^{0.40}$ both for marine soarers and for land soarers (figure 8.8). Heating and cooling time constants in water for geometrically similar animals made of the same materials would be proportional to $(\text{body mass})^{0.67}$, and the observed time constants for different sized reptiles are proportional to $(\text{body mass})^{0.67}$ and $(\text{body mass})^{0.71}$ (figure 7.4). The deviations from geometric similarity modify the effect of differences of size but do not cancel the general trends.

These points form part of the message of this book, but there is another very important part: engineering theory, designed for application to man-made structures, can also help us to understand the structure and behavior of animals.

I used an idea that had its origin in shipbuilding to define equivalent running speeds for animals of different sizes, and information from soil mechanics to assess the danger of dinosaurs getting bogged down in mud or sand (chapter 3). I used methods developed by engineers to calculate stresses in structures such as bridges when I discussed how athletic dinosaurs could have been (chapter 4), and again in calculations about dinosaur neck ligaments (chapter 5). The calculation of forces for a collision of dome-headed dinosaurs could have been applied to automobiles, and the discussion of *Parasaurolophus'* voice used simple acoustics (chapter 6). The discussion of dinosaur heat balance (chapter 7) used theory such as heat engineers use when designing central heating systems. I used aerodynamics developed for application to aircraft to show how pterosaurs probably flew (chapter 8) and similar theory applied to water instead of air to show how ichthyosaurs and plesiosaurs may have swum (chapter 9). Geophysical information was needed in the discussion of dinosaur extinction (chapter 10) and diffusion theory in the discussion of birds' eggs (chapter 11).

Physics and engineering are as useful in the study of living animals and of the human body, as in the study of dinosaurs and other extinct animals. Physics is the basic science of matter and energy, and engineering is physics applied to structures and machines. They and chemistry are the sciences that biologists need to explain the structure and mechanisms of living things.

INDEX

(italics indicate figures or tables)